D1537818

BLINDSIDED
Planet X Passes
In 2003
EARTHCHANGES!

Copyright © 2001 by Mark Hazlewood
All rights reserved. Printed in the United States of America.
No part of this publication may be reproduced, stored in
a retrieval system, or transmitted, in any form or by any
means, electronic, mechanical, photocopying, recording,
or otherwise, without the prior written permission of the author.

ISBN
1-931743-40-1

Library of Congress Cataloging in Publication Data
2001099807

Mark Hazlewood
Blindsided

10 9 8 7 6 5 4 3 2

FIRSTPUBLISH
A Division of the Brekel Group, Inc.
300 Sunport Lane
Orlando, FL 32809
407-240-1414
www.firstpublish.com

Forward

Why do I believe Planet X will be returning in 2003?

A combination of in-depth research and life experience. In the mid-'70's, at age 19, I started work with a solar cell corporation. It had technology that could undercut the cost of fossil fuels and nuclear power. Twelve years later it became clear that there were people that controlled the energy cartel, banks and media that were engaged in a combination of underhanded financial and legal maneuvers to stop us. I wasn't looking to find that these controllers existed. I had no interest in them. I did become acutely aware of them first hand, having watched as they slapped the company I worked with down and effectively shelved the technology.

Fast Forward>>

Over the years, I've had a casual interest in prophecy relating to earth changes. I read perhaps 100 plus differing intuitives' works from various cultures and times around the globe. Many curiously coincided on what was expected and had an approximate time frame of around the turn of the millennium for their occurrence. After concluding that there was a high probability that this was not a coincidence, I figured that earth changes were possibly part of our immediate future. Because I had no indication of why this would happen or exactly when, I temporarily let it go and lost interest. Prophecy alone did not give me what I was seeking, to be sure.

Over the years, I've worked on my dietary habits and studied and practiced yoga off and on. This has had an incremental awakening on my spiritual connection. I know I've a long way to go; at the same time, I see subtleties today I wouldn't have dreamed existed decades ago. For now I'll keep these insights mainly private.

I've an adult son and a grandson. In 1998, about 3 years ago, my son and I both had simultaneous earth change dreams of almost the exact same nature. We discussed it, and I decided to probe my spiritual connection for an answer as to the why of these dreams. I wasn't given anything. However, within 36 hours, by my own probing efforts, I had stumbled onto that answer.

I discovered that there was a thread in my study of prophecy and intuitives that I'd dropped several years prior, which I'd missed, about a rogue planet, dark star or passing comet. This began my research of Planet X. Along the way, I found an overwhelming amount of archeological evidence that coincided with X's orbit. Earth had been speaking very loudly on a cyclical basis as to when she was experiencing calamities. The majority of archeologists and scientists hadn't studied the ancient historical accounts of Planet X and found the regular approximate 3600 year orbit to fit the pieces of the puzzle together. The ice age phenomenon was real but given the wrong name and reasons for its occurrence. From a wide variety of ancient and modern cultures, I found several names for Planet X. I've 34 names for X in this text.

Along the way, I discovered the approximate 7-year period of non-religious tribulations connected to X. This is where, during the dense brown dwarf's approach, its far-reaching electromagnetic and gravitational arms start to slow Earth's rotation while heating up and changing its core flows. This is the reason for much that has happened in the last 5 years. This is the cause, since around 1995, of the worldwide mega weather changes, droughts, fires, El Niño, La Niña, the sudden doubling in deep quakes starting in late '96, the up-tick in mega-quakes of 7.0 or greater in the last two years, quake swarms, domino quakes and the increase in volcanism, among other changes.

The media controllers are falsely blaming widely known and discussed weather changes on global warming or sun cycles. They are keeping quiet about the same level of abrupt changes in quake or volcanic activity happening simultaneously, although they can't help but report major events hitting populated areas. It'd be quite a stretch for them to blame all this on global warming and sun cycles too.

NASA discovered and excitedly disclosed having found Planet X. It was the last day of 1983 that announcement was made by the chief scientist of the IRAS satellite to 6 dailies. I've got all the headlines in my book. Shortly thereafter, the media controllers got wind of the disclosure and clamped down by having them retract the statements. They've since made up an extremely simplistic story to explain away the original perturbations of the outer planets that ultimately led to finding Planet X. As far as explaining away having made the announcement, the best they could come up with is that it was misunderstood.

Only the largest controlled scopes could have seen Planet X until early this year. It was easy to cover up until then. Small to mid-level observatories around the world kept spotting it in early 2001. To quash the concerns, the controllers announced they had found the "Largest Asteroid in history August, 2001."

This asteroid is one of many objects around the area where Planet X is approaching that are being identified. These disclosures and techniques are meant to confuse, pacify and redirect the publics attention away from the brown dwarf.

Occasionally you'd expect, with such an overwhelmingly important worldwide story, that even has intentional leaks, some details would slip out from a few national leaders in the know. Russia's leaders did just that.

Here are quotes from them, sent over Europe's Reuters news service wire Sept. 13th, 2000: "Anticipated Chain of Disasters Due to Hit in 2003" and "Massive Population Shrinkage." They also wondered out loud, on recorded microphone during a parliamentary meeting, if Russia would even exist as a country afterwards.

Zecharia Sitchin carries weight in the scientific archeological community and has a degree in journalism. The enforcer controllers made clear to him what he was allowed and not allowed to say and print early on. Sitchin's colleague astronomer

Dr. Harrington of the United States Naval Observatory met with an early demise. This happened because of his adamant open disclosures to the media about having found the 10th planet. A loud message was sent to everyone with weight in the scientific community or credible insiders in the know. If you value your life, keep quiet!

The story has another strange twist. Subgroups of these same controllers I keep referring to have their own quiet campaign to get the knowledge out through grassroots sources. I fit perfectly in that category, having no scientific degrees to speak of nor weight in the scientific community.

They don't want panic or the collapse of the financial markets to start too soon. I don't either, but my conscience drove me to the edge until I wrote this book. I'm sure you can imagine these controllers have a need to relieve some of their guilt for having held back the information and having cost millions if not billions of lives. Of course they just may want a few more to survive for selfish reasons. This would be so they have plenty to enslave for farming purposes, around their already stocked fortified enclaves, once they begin to run low a few years from now.

I've heard now that Sitchin is being more forthcoming with the fact that he knows X is very close. He is now occasionally saying that **"the time for Planet X's or Nibiru's return is now"**. I guess his concern for his personal safety has finally taken a back seat to the lives he could be saving by speaking up.

A note about the tragic World Trade Center and Pentagon destruction. We've been BLINDSIDED by these events. There may be some positive changes in consciousness that have come about as a result. It now should be obvious there's one direction to ultimately look in order to protect yourself and loved ones, in the mirror. Government officials can not be trusted to protect us nor give us the information needed in order to protect ourselves. If you've ever doubted it before, you now clearly understand our way of life can and will be completely changed with one day's passing.

Six to eight weeks prior to the passing of Planet X, all will be able to see it in the sky as sort of a red cross with their naked eyes.

Will you be BLINDSIDED by the passing of Planet X in 2003?

Table of Contents

Earth Science Evidence of Abrupt Regular Earth Changes From The Approximate 3600 Year Orbit Of Planet X

Government Wraps Things Up By 2003

BLINDSIDED

Warning, stop right here! Read no further if you don't want to understand how, why and when the earth's surface is going to change so abruptly, as to possibly leading to taking out as much as 90% of the population in many areas. You'll find why these changes happen regularly, how often, who knows it, and how the information is being given out. You'll also grasp why this may be your first exposure to this knowledge, or at least the first untainted divulging.

There's absolutely nothing anyone can do about it. Our sun's 10th Planet disrupts the surface of every planet in our solar system, as it passes on its regular orbit. Earth is not being singled out. Only a few hundred million people will survive. Now ask yourself this question. Do you want to be one of them? Do you have a sincere desire to pick up the pieces and help build a new one?

This world's culture has unhealthy and healthy of all ages. Are you healthy and mature enough mentally, physically, emotionally and spiritually to give it a go? If not, forget you ever read this. The details and sources will be worthless to you in the short life you have left. Just keep living from day to day like there's no tomorrow. If you've any unhealthy personal habits like drug use, lack of exercise or over eating, don't worry about changing them. Most won't live any longer doing so. So enjoy whatever small pleasures guilt free from this point on. The things you've been putting off, do them now. You may never get another chance. If you don't like doing something, stop it. There's no point anymore.

Now for the rest of you who want a tomorrow, to be informed, forewarned and attempt to place a reservation for yourself and loved ones in the aftertimes, this could be the hardest hitting book you'll ever read. The controllers of the media, money markets and the world's government leaders are fully aware and prepared. They are not going to announce the impending calamities, although Russia's leaders did pretty much let the story slip out at least once, which I'll detail a little later on here. For the most part they're only allowing this information out through grass-roots sources. So escape with me here among the grass, and you'll find out the many roots this information is being brought to you from.

Keeping this story out of the major media will delay the collapse of the world's economy, stock markets and real estate prices until the very last days costing billions of lives. From this point on, don't expect the price of energy to come down much. These corporate energy controllers are fully aware that this is their last hurrah to gouge money from you.

I am not the origin for the information that follows. I have simply gathered together several sources for your perusal. I did not discriminate from where I obtained the information. Part of my motivation is to get you directly involved in looking inward to your own gut feelings and doing your own research. Many already sense something is wrong big time with Earth but can't put their finger on it.

Be careful how you view this material. Some tend to ignore all sources in a body of information, that causes great consternation, when only one reference isn't acceptable to them. This kind of thinking is illogical. The utmost seriousness has gone into putting together this work of warning. A myriad of sources from the sciences, government, history, politics, legend and spirit is combined here. You will do yourself and loved ones a great disservice if your mind set is to discount all <u>you consider credible</u> when presented side by side with sources your cultural experience and educational programming have not deemed worthy of consideration.

Before the time of the greatest events, the overwhelmingly vast majority will see, hear and experience enough to know that our earth is about to change dramatically again. By the time your awareness peaks into realization, it may be too late to prepare if you don't begin to get up to speed on the situation now. I've spent my time researching, accumulating, organizing, discussing, and weeding out side agendas. I present this subject in such a way as to bring a quick clarity to the layman or someone who is completely unfamiliar with it. If you've never read anything about the when and why for earth changes before, this work may be all you'll ever need. Reading up on endless conspiracies, for the most part, will not help you with your life. Understanding only this one may be important for your survival. Nailing down the science, that shows the regularity of these events is important to see clearly how close to the end of this cycle we are.

Questions will come to mind when the acceptance of what is about to happen does sets in. Will I have time to make it to a safe area? Am I there already? Will I have resources enough to sustain myself over an extended period of time, or the ability to create needed sustenance on an ongoing basis? Do I care to leave all which I hold dear to start anew?

<u>Am I just too married to or fearful of leaving my things, lifestyle and community, even at the expense of my life, if I can't take them with me??</u>

2

Watch your thinking carefully as you read. Your mind may make up unsound reasons for ignoring the obvious to avoid the discomfort of having to prepare.

It is an honor to be a vehicle or messenger of this information. I'm one of the many that have chosen or have been chosen or some combination thereof, to do this work. I'm sure I'll meet some people in the aftertimes that acknowledge me for warning and helping them survive one of the most severe physical calamities Earth has gone through. On the other hand, others may wish they never learned or prepared because of the difficulties presented afterwards, compared to their lifestyle before. At that time I would simply say that YOU were primarily responsible for drawing the information toward yourself and determined your level of personal preparation. I'm not fishing now for your future thanks, but am trying to avoid you berating the messenger. Just learning now and not being able to prepare because of family ties or financial binds could end up causing needless worry, prior to your passing on. <u>So, be careful before you jump in too deep here. It's not always better to know.</u>

My belief system tells me that there is a reason for everything. Science is finding greater evidence for order in chaos at every turn. For those who would label me a messenger of doom, please do you and me a favor; DON'T READ THIS BOOK. My focus is to find people who want this information, or who have been drawn to it by whatever means to make their right choice. Staying alive might be a good one. The fact that you're reading this perhaps means that your gut instinct is telling you something is not quite right with the world, and the reasons you've run across thus far have been less than adequate. <u>There's no doom in being made aware that you live on the railroad tracks, the train is approaching and what time it will arrive, so you can pull up your roots to get the hell out of the way. It's ignorance that will put you in harm's way during the up and coming cataclysms. There will be many places that have a good chance of being safe.</u>

I'm now going to paint a word portrait using some of my words and mostly those of others. In order to view the mosaic, look at it in its entirety. Focusing too closely on one part or another might result in your missing the big picture. This work could have been 10,000 pages or more. There is much more than that volume of material confirming all that is presented here. Studying the data and the myriad of different sources becomes very repetitive. I strongly urge you to do your own research if you need more information. Much more related truth is out there. Since I've outlined where to look and what to look for, it will be easy to obtain more sources for this subject matter. If you're like me, once you've seen enough of it, you'll quit looking for more, even though you've come to the understanding so much more exists. You'll then make your plans and go back to living your daily life. <u>On the back burner of your mind you'll know there are some wild times and big changes just around the corner. Plus, you'll know, more than likely, that you'll survive them because of your ace in the hole of being informed.</u> Are you getting my drift here? These events are going to be exhilarating, adventurous and monumental. Doom, gloom and fear shouldn't be your mind-set whatsoever. The most exciting, serious, unbelievably dramatic scenes you've ever viewed at the movies will be played out in your real life from day to day shortly, whether you watch the previews of this book or not.

Observatory Sightings of Planet X In Early 2001!

****

In 1983 X was first spotted by the IRAS (Infrared Astronomical Satellite), since then there has been a complete blackout of its existence by the controllers of the media and major observatories. X is now too close to cover up from smaller scopes. For the first time in modern history, in the year 2001, sightings of Planet X are being openly recorded. Three separate sightings from different parts of the world have reported it. The three observatories were located in South America, Switzerland and Arizona. I include two of the sightings and show you where to find the third and future ones later. Here is a short report from the Lowell Observatory in Arizona.

April 04, 2001, 03:06:45, Wed: The operator described the object as diffuse and of approximate magnitude 11. The coordinates (in degree/ minute/ second format), with a margin of error according to the operator of +/- 20 seconds (about .006 degrees) RA and +/- 10 seconds (about .003 degrees) Dec were: RA 05 09 09Dec +16 31 49
In degree format the coordinates are: RA 5.1525Dec +16.5303

Next is an excited confirmed report and then denial from Switzerland.
Subject: 12th planet discovery.
Date: Wed, 07 Feb 2001 23:30:32 +0100
A whole team was contacting every observatory in France — just sent a message. The Neuchatel observatory got it. They are very excited, wondering if it is a comet or a brown dwarf, through the latest coordinates given. The daughter of the astronomer reports that they suspect a comet or a brown dwarf on the process to becoming a pulsar since it emits "waves." For those who would read French, I copy the message below: Salut!
Bon, les jeunes y'a du nouveau. J'ai voye les donnees concernant la 12e planete a une amie, et voici ce qu'elle me repond: L'observatoire de châtel (celui du pater-nel) toute première réponses: oui, ce pourrait être une comète. Elle est sur un des

bras d'Orion(?) et vont se mettre à mieux regarder pour valider ou non "la naine une"...car je ne sais pas si tu sais, mais ce stade est juste avant celui du pulsar et donc émet des ondes... CQFD... je me demande ce que la Terre en reçoit ou en recevra mais... Mystère et boules de gommes... Attendons les autres labo... mais celui-ci en particulier je lui fait confiance car il ne jouerait pas la carte du complot avec mon père... ça franchement non.. (autant dire que le — dit Père est tout exité!!!)

We were told by these excited folks who first sighted it, that we would get "at least" an image; next there was a long silence for over a week. Afterwards came the official denial that such a sighting had ever occurred at Neuchatal. As in most cover-ups, believe what is said first, and not what is retracted later. So they possess an image of X.

****

Later, I'll point out where the coordinates for the three sightings came from, how to find future coordinates, and where you can read about all the curious blocking attempts that happened before the third sighting.

The powers that control the major observatories have been quietly observing it for years now. Because X is so close and easily observable, the only thing left for the controllers to do now is to discount what it is. When the fact of X's existence becomes more widespread, the powers that be will be used to drown out the reality of the calamities with words of reassurance. Their power to overshadow the truth is vast. Do expect the cover-up to continue to work for a short while with most.

For the better part of last century, astronomers have noticed perturbations in the outermost planets. These perturbations could only have been caused by another large heavenly body that is part of our solar system. Finding the archeological evidence is simple. The more you dig for evidence, the more you will find of quick regular severe surface changes and pole shifts at every turn. I have more than a sufficient quantity of these studies for you to browse through right here and point towards where to find several others that have accumulated much more. Later I'll show how the Ice Ages fit into the puzzle.

The 12 Planet by Zacharia Sitchin

Turning to ancient history will put the evidence into context. Many past civilizations explain Planet X's orbit length and describe the destruction it causes during its passage in eloquent detail! Start with "The 12 Planet" by Zacharia Sitchin.

5

Even historians of Astronomy are aware of the legends of a so-called "Star or Comet of Doom" or "Nemesis," that brings with it debris, meteors, and upheavals.

For the most part aware political leaders have no clue what to do about the masses, so of course they aren't doing much. Their attitude is that an honest announcement would cause more problems than not. Writing this book to help and inform others is where I stand.

Finding out what changes Earth goes through and when the changes start, relative to each passage, will show you how many years we've been influenced by X's current approach and when it will arrive. This is not rocket science. It's easy to understand, but sweeping earth changes in your immediate lifetime are difficult to fathom because of the enormity.

Since 1995-6, Earth's weather has changed dramatically. This is how it starts every time! To understand what the current effects of X are now, weather and seismic activity should be closely monitored throughout the globe. Watching and listening closely to the intentional and unintentional warnings from the world's government officials will add clarity to the picture. Studying legend, folklore and prophecy might finally grab your heart and seal your knowing. Taking a little vacation time to reserve some time at a small observatory with coordinates and dates to find X, that I'll show you how to get later, might just shock you into action once you've seen it with your own eyes.

Now, if you've some extra money, you could pay others to do your footwork for you in terms of further research and observatory time. I do caution you that there is as much disinformation as correct information. Much of the same truth I present here is intentionally mixed with skepticism and fabrications elsewhere. If you don't do your own research, how do you know which is being reported back to you? At some point you'll have to get used to doing things for yourself and possibly for others in the aftertimes.

The monetary playing field will be equalized after X's passage. For the first few years, cash will be worthless. Even right preceding to the events cash could devalue severely. The longer you wait to prepare, the greater the price you'll have to pay. If you wait too long, you'll pay the ultimate price.

In the aftertimes, barter and co-operation will be king if you luckily end up in the right community. Mad Max scenarios will reign elsewhere for a while, until they play themselves out.

At a certain point you will have learned enough to make a prudent decision to leave unsafe areas. Starting with just a search of Pole Shift or Earth Changes information will land you in the middle of a mountain of evidence confirming Earth's surface changes regularly and quickly.

Ice Ages — Grade School Science

<u>Earth truly is one of the most dangerous planets in the Universe.</u>
Remember back in grade school, when you were first taught about the Ice Ages? This was your first big hint that something goes wrong with your home planet Earth on a regular basis. Most of the researchers, who all agreed this phenomenon was real through the science of archeology, just didn't figure out the correct reason for it, or exactly how often it happens and why. Plus, they really mislabeled it. The poles and ice shifting to different parts of the globe are certainly not the only thing that happens. Different areas of the planet would show varying times for the last event when looking for only the effects of ice and cold. Yes, they discovered something absolutely real. No, most of them didn't know the correct why for the phenomenon.

The so-called Ice Ages are thought to have happened over a period of many years. In fact, the great mammoths found with tropical food in their stomachs that were flash frozen is just one fact from the tip of the iceberg of evidence that shows changes happen very quickly. A little later I'll bore you to tears with Earth Science to show this. There are no so called ice ages. Earth's crust slips over its molten core periodically as Planet X passes. This shifts the poles to different areas of the world periodically. Glaciers don't just magically start moving up or down from Earth's current poles on a regular basis. The poles themselves shift to different parts of the globe in a matter of hours. All life where the new poles settle flash freezes instantaneously. A few thousand years ago Greenland was a polar ice cap of Earth. There's still way too much ice there for its latitude relative to the current poles.

You are solely responsible for determining for yourself what is proof or evidence. Even though this is very simple, most do not possess the mental capacity to piece together the varied disciplines to see what's right around the corner, even when laid out as concisely as I do in this short work. If they do see it easily, there's a shorter supply of people that are type "A" individuals that take action in their lives. Most will not want to leave obvious unsafe areas until Planet X is squarely in their view. In this case that will be fatal for the majority. This is just the sort of realization that gets to me. Even at my best I can only expect to awaken a small minority. For those who read, grasp the situation and prepare, you are in a very exclusive club. I don't like being correct on this issue.

Warnings will not be broadcast by public officials over the TV, radio, or newspapers. <u>Warnings will not be broadcast by public officials over the TV or newspapers.</u> *Warnings will not be broadcast by public officials over the TV.* A knock on the door from government workers of your city or state, saying it is time to evacuate, will not be forthcoming. There are just too many people to deal with. Don't you understand? There would be no place prepared to evacuate to, or resources available to sustain the many over any length of time. Consequently the call to leave will not come. It's just too huge. You are alone in determining your fate. For those of you who tend to ignore important issues or can't handle the truth, this is one truth that will find you. Will it be when you look over your shoulder and see a tidal wave or a building collapsing down on you or in time enough to reach a safe area?

Remember this if you choose not to move from <u>unsafe areas such as within 100-200+ miles of any coast line. This includes all of California and Florida.</u> Please pay attention to everyone around you as the time approaches if you haven't made your move. You will see families silently moving away with talk of just a vacation, covering up that they are scurrying away to their shelters. Now for those who think God will give you a warning and take care of you. Hold that thought and consider this written work may be <u>your warning.</u>

Arctic may not have been icy in Ice Age
Friday, September 07, 2001
By Reuters

LONDON — Imagine the Arctic Circle in the Ice Age. White and cold are probably two of the adjectives that first spring to mind. But archeologists say a recent discovery of animal bones and stone tools means humans lived there more than 40,000 years ago, and the region then may not have been covered in ice at all.

An international team of scientists said this week they had discovered stone artifacts; reindeer, wolf, and horse bones; and a mammoth tusk with human-made marks on a dig at Mamontovaya Kurya in the northern reaches of Russia. The finds are the oldest documented evidence of human activity so far north.

"The bones and artifacts found suggest that the northeast must have been relatively dry and ice free in this period of the Ice Age," archeologist John Gowlett said, commenting on the research in the scientific journal Nature. "It is not possible to determine whether they were left by Neanderthals or by some of the first modern humans in Europe, but ... knowing who made the tools is less important than simply knowing that someone was adapted to the cold conditions," said Gowlett from the University of Liverpool.

Scientists Pavel Pavlov and John Inge Svendson, who made the discoveries, said they were unsure who the mystery Arctic inhabitants were.

"Either the Neanderthals expanded much further north than previously thought, or modern humans were present in the Arctic only a few thousand years after their first appearance in Europe," they said.

Sumerian Descriptions Of Our Solar System

The 6,000 year old Sumerian descriptions of our solar system include one more planet they called "Nibiru," which translates into "Planet of the crossing." The

8

descriptions of this planet by the Sumerians match precisely the specifications of "Planet X" (The Tenth Planet). Views from modern and ancient astronomy, which both suggest a highly elliptical, comet-like orbit, take Planet X into the depths of space, well beyond the orbit of Pluto. The Sumerian descriptions are being confirmed with modern advances in science. There are actual diagrams on well-preserved tablets from the Sumerian times that show how their accuracy for describing the planets is overwhelming!

United States Naval Observatory Calculations

Recent calculations by the United States Naval Observatory has confirmed the orbital perturbation exhibited by Uranus and Neptune, which Dr. Thomas C. Van Flandern, an astronomer at the observatory, says could only be explained by "a single planet." He and a colleague, Dr. Richard Harrington, calculated that the 10th planet to be two to five times larger than Earth, and it has a highly elliptical orbit that takes it some 5 billion miles beyond that of Pluto.

In January, 1981, several daily newspapers stated that Pluto's orbit indicates that Planet X exists. The report stated that an astronomer from the U.S. Naval Observatory told a meeting of the American Astronomical Society, that irregularity in the orbit of Pluto indicates that the solar system contains a 10th planet. He also noted that this came to no surprise to Zecharia Sitchin, whose book came out three years prior.

I've corresponded with one who has spoken with Sitchin in a private group. Sitchin knows that X is very close, but because of his position in the scientific community — well, you figure it out! I'm glad I don't have a Ph.D in front of my last name or hold weight in the scientific community. Sitchin needs to concentrate on selling more books. He's too hot to talk about what he knows.

In 1982, NASA themselves officially recognized the existence of Planet X with an announcement, "An object is really there far beyond the outermost planets." Today NASA is not being forthcoming about X.

New York Times June 19, 1982

Something out there beyond the furthest reaches of the known solar system is tugging at Uranus and Neptune. A gravitational force keeps perturbing the two giant planets, causing irregularities in their orbits. The force suggests a presence far away and unseen, a large object, the long-sought Planet X.

There are mathematical irregularities in the orbits of the outer planets. Astronomers are so certain of this planet's existence that they have already named it "Planet X— the 10th Planet."

Headline News — Planet X Has Been Sighted! Chief IRAS Scientist JPL December 30th, 1983

One year later in 1983, the newly launched IRAS (Infrared Astronomical Satellite) quickly found Planet X. This is a summary from the Washington Post from the chief IRAS scientist of JPL in California: "A heavenly body as large as Jupiter and part of this solar system has been found in the direction of the constellation of Orion by an orbiting telescope." Now read that again. The disinformation drain brain crew did their best to try to rewrite history again after this announcement was made. The public at large must be kept in the dumb and dumber category is their thinking. They can't handle the truth.

The telescope found it right where it was sent to look. These scientists had known of its existence and location for years, but wanted to confirm it with our own technological eyes. This is a fact I strongly urge you to check out for yourself. There have been attempts to cover up this event and rewrite history. Headlines from a few other dailies read as follows; "Mystery Body Found in Space." "Giant Object Mystifies Astronomers." "At Solar System's Edge Giant Object is a Mystery." "When IRAS scientists first saw the mystery body, they calculated that it could be as close as 50 billion miles and moving towards earth."

Tombaugh was given credit for discovering Pluto in 1930, although Lowell spotted it earlier. Christie, of the U.S. Naval Observatory, discovered Charon, Pluto's moon, in 1978. The characteristics of Pluto derivable from the nature of Charon demonstrated that there must still be a large planet undiscovered because Pluto could not be the cause of the residuals, the "wobbles" in the orbital paths of Uranus and Neptune clearly identifiable. The IRAS (Infrared Astronomical Satellite), during '83-'84, produced observations of a tenth planet so robust that one of the astronomers on the project said that "all that remains is to name it" — from which point the information has become curiously guarded.

In 1992, Harrington and Van Flandern of the Naval Observatory, working with all the information they had at hand, published their findings and opinion that there is indeed, a tenth planet, even calling it an "intruder" planet. Andersen of JPL later publicly expressed his belief that it could possibly be verified any time. The search was narrowed to the southern skies, below the ecliptic. Harrington invited Sitchin, having read his book and translations of the *Enuma Elish*, to a meeting at his office. They correlated the current findings with the ancient records. Harrington acknowledged the detail of the ancient records while indicating where the tenth planet is in the solar system.

It is the opinion of this author that, in light of the evidence already obtained through the use of the Pioneer 10, 11, the two Voyager space craft, the Infrared Imaging Satellite (IRAS, `83-84) and the data available to Harrington when consulting with Sitchin, that the search has already been accomplished. In fact, the planet has already been found. It is interesting that Harrington dispatched an appropriate telescope to Black Birch, New Zealand to get a visual confirmation. The data lead him to expect that it would be below the ecliptic in the southern skies at this point in its orbit. On Harrington's early death, the scope was immediately called back. Hmmm — as one observer noted, "almost before he was cold."... Robert Harrington used to be the head of the Naval Observatory, and Tom van Flandern worked closely with Harrington at the US Naval Observatory. (*The Alien Question: An Expanded Perspective,* by Neil Freer)

What obvious message do you think was sent to Sitchin, Van Flandern and anyone else in the know, when Harrington suddenly met with an early death at the same time the scope was being pulled back? Sometimes a so-called accidental death is meant to accomplish more than just keeping one person quiet.

Nemesis Theory Nemesis Fact

In 1985, numerous astronomers were intrigued with the "Nemesis Theory." This was proposed most recently by Walter Alverez of the University of California and his father, the Nobel prize winning physicist Luis Alvarez. They noticed regular extinctions of various species (including the dinosaurs), and proposed that a comet, "Death Star," or planet periodically brings with it a shower of meteors and smaller comet like objects that wreak havoc, death and destruction to the inner Solar System, including Earth.

In August, 1988, a report by Dr. Robert S. Harrington of the U.S. Naval Laboratory calculated that its mass is probably four times that of Earth.

Planet X is 4 to 5 times larger than of earth, 20-25 times its mass and nearly 100 times as dense. X is a professional wrestler of the planetary community compared to other planets of our solar system. It is a slow smoldering brown dwarf. When X passes between earth and our sun, earth will align to its strong magnetic or gravitation temporarily, instead of our sun's. For a short while X will be earth's strongest gravitational voice or influence. Earth's rotation will then pause for a few days, like it has many times before. In ancient historical text this is commonly regarded as myth. The controllers of NASA are fully aware of its truth.

Planet X's destruction and disruption of Earth should change your view of ancient history. The ancients were not the unsophisticated people common history books have led you to believe. Earth societies have reached a pinnacle of development several times before only to be put back in technological and evolutionary time by the wrecking ball of X. Every time the ones left to pick up the pieces and start building over were so busy surviving for the first few decades afterwards, that the elevated cultures they originated from disappeared into myth and folklore. Many of the advances and discoveries of our so-called modern world took centuries to be <u>rediscovered</u> anew. Listen closely to whatever texts and knowledge from our ancient ancestors that still remain.
<u>Their truth is our truth.</u>

Thanks to 2 years of research by Andy Lloyd, another piece to this puzzle emerges. Nibiru is one of many names from ancient culture that reference Planet X. A segment of Andy's research follows:

"The unstable nature of the dark star's orbit means that it has precipitated periodic change to the orbital radii and climates of the terrestrial planets over the last 4 billion years. Among my findings is that Nibiru is a brown dwarf, a failed star capable of emitting only the faintest reddish light, but whose gravity and infra-red energy emission is sufficient to warm its habitable moons. These moons number 7, it seems, and the infrequent perihelion passages of Nibiru are associated with the 'return of the gods'. When Nibiru is in close proximity to the Sun, both these bodies become excited by each other. The brown dwarf, in particular, becomes 're-lit' from its slumbering embers and 'flares up' with reddish light. The coronal discharge emitted is then swept back by the action of the Solar Wind, giving the appearance of fiery wings whose flight is directed towards the Sun."

34 Names for Planet X

Ancient history, astronomy, folklore and prophecy record many names for Planet X. The Sumerians called it the "<u>12th planet</u>" or "<u>Nibiru</u>" (translation; planet of passing). Between the Babylonians and Mesopotamians there were at least three names: "<u>Marduk</u>," "<u>The King of The Heavens</u>," and The "<u>Great Heavenly Body</u>." The ancient Hebrews referred to it as the "<u>Winged Globe</u>" because of its long orbit high among the stars. The Egyptians had two names "<u>Apep</u>" or "<u>Seth</u>." The Greeks called it "<u>Typhon</u>" after a feared leader and "<u>Nemesis</u>" (one of its most telling names). Other ancient peoples have given it labels such as; "<u>The Celestial

Lord Shiva" and "God of Destruction." Ancient Chinese knew it by: "Gung-gung," "The Great Black," or "Red Dragon." The Phoenicians said it was "The Great Phoenix." The Hebrews called it "Yahweh." The Mayans called it "Celestial Quetzalcoatl." The celestial body was known to the Latins as "Lucifer." The Russians give the name "Great Star." Revelation (8:10-12). The name of the star is "Wormwood."

From works of prophecy, there are other names for X. The "Red or Blue Star" is of the Hopi Indian and Gordan Michael Scallion designation. The "Fiery Messenger" is in the Ramala prophecy. The "Great Star" is from the Book Revelation. "His Star" is how it is referred to in Edgar Cayce Readings. The "Great Comet" and "The Comet of Doom" is right out of the Grail Message. From an early English prophet named "Mother Shipton," "The Fiery Dragon" was the name she gave, as seen from her second sight. Our solar system's "10th Planet" is "X."

From modern astronomy there's: "The Intruder," "The Perturber," and lastly the 1977 Chileans' discovery labeled "Bernar-1." Apparently 25% of all observed comet orbits are being measurably perturbed by the magnetic or gravitational pull of this planet. The largest of our planets' orbits are being perturbed toward Orion.

No matter what the name used, it's the same object that has the same effects before and during its passage of earth. The Sumerians also had a name for its approximate 3600 year orbit, "A Shar." The ancient Hindu astronomers gave the name "Treta Yuga" for its 3600 year orbit. The destruction X causes even had a label "Kali Yuga." Some of the names given by the ancients are akin to names for God or the devil. These people viewed the power and destruction that this planet brings with it to be so significant as to believe it could only come from the hand of God. It's not that they worshipped this planet, rather they just had great respect for the sweeping changes it brought with each passing.

Discovering Archeology, **July / August 1999**

Look to *"Discovering Archeology,"* July / August 1999, page 72. The date 1628 B.C. is given for a world wide catastrophic event. The planet wide effect shows up in the growth of trees that can be viewed by studying the narrowed rings from that time. This is approximately 3600 years ago and coincides with Nibiru's return at this time. Page 70 shows a medieval picture with a large comet-looking object, appearing as big or larger than the sun streaking across the sky horizontally with a giant tail. This is noteworthy because of its comparative size and direction in the sky. It is not headed down toward earth and yet is still causing much destruction below. The presupposed premise of the article is that a comet impact had to have taken place to cause the global calamities, and yet the very picture chosen from medieval times denotes an object just passing by. Pictured under this massive object moving overhead is a town that is shaking apart with hysterical and even some suicidal people in the streets.

Medieval Drawing

Destructive Brown Dwarf

Planet X's orbit takes it back and forth between two suns. The other sun it orbits around is our sun's twin. Now, you might think that our solar system is not binary, although most solar systems have been recently discovered to be binary. Our sun does have a twin. It lies in the direction of Orion. It is not composed of the same material as our sun and thus will never ignite. It is a dark or dead sun and does not revolve around our sun like most twin suns do. Why hasn't this discovery been announced you might ask? It has if you know where to look. It's real simple why the information isn't more widely known. There's a brown dwarf star that orbits between both suns, instead of around just one of them. Its orbit is rather long, being about 3660 years, so we don't see it go by too often. It is still one of our family of planets in our solar system. It's like a half brother that's been ostracized by it's two parents, but still comes visiting habitually. X is kinda like the black sheep of our solar system's family. When it finally comes by for a visit it just screws everything up and then leaves. It happens every time, so I guess the rep is well deserved.

Now stepping back away from humor, if you're going to try to keep a lid on the "destructive brown dwarf" (nothing kinky here) that's currently inbound, it's not a very good idea to notify John Q. Public about the relatively recently discovered second sun that's one of the foci of its orbit. There's no reason to promote important clues if you don't have to. Besides, only very powerful and expensive scopes or deep space probes can see the other sun. This makes it fairly easy to keep it under wraps from the majority. This is one of many reasons the orbiting Hubble telescope live feed is hidden from our view. Hubble was paid for with tax dollars, so you'd think they would give us a live peak. Nope, it's not going to happen.

Diagram Of Our Solar System Includes The Dead Twin Sun and the 10th Planet a.k.a. Planet X

This diagram appeared in the 1987 edition of the *"New Science and Invention Encyclopedia"* published by H.S. Stuttman, Westport, Connecticut, USA. The continuing conspiracy to cover up X falls short when matter of fact evidence is found in respected credible publications like this. The article accompanying this diagram was discussing the purpose of the Pioneer 10 and 11 space probes. Clearly shown on the diagram are our suns dark twin Dead Star and Planet X, a.k.a. the 10th or 12th Planet. The primary point made was to show the paths of the two probes and how they created a triangulated sighting. The probes just happened to be verifying, through this triangulation, the sighting of Planet X. The history of deep space probes is mentioned because Pioneer 10 and 11 were said to be one of the first.

There's absolutely no mention of any controversy about what was out there or what the probes were looking at. The disinformation to hide X's approach and long known history of devastation it causes as it passes each time, did not reach this far. It's not hard to imagine a conspiracy as large as this one, just can't weed out all the evidence. It's a big world with a lot of people, publications, and information sources, present and past.

You know what could be a great thing about this cover-up? The information that it misses may get a bright light of emotion and attention shone on it, that wouldn't normally have been there without the effort of trying to hide it. There's a possibility if this information I present gets out quickly and broadly enough, there might be more people paying attention to X than if there was never any conspiracy to cover it up at all. That thought is just a hope and a prayer at this point.

Encyclopedia Diagram

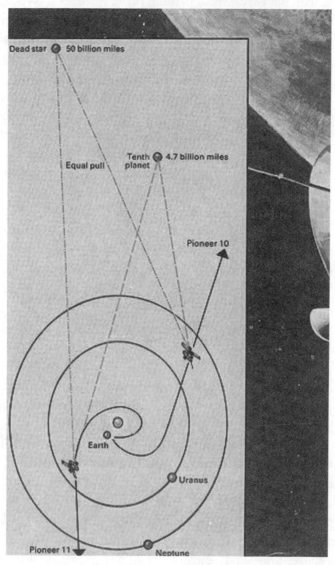

When a planet's orbit is between two suns, instead of one, that orbit is a bit peculiar. It spends 99.99% of its time slowly going away from one of the suns, after it passes it initially quickly, to reach the half way point. Then, as if falling like a rock off a cliff, the gravity of the sun it's approaching takes over, and in a relative flash it travels the other half of its journey. X reached the halfway point, after a little over 1800 years of traveling away from the Dark Sun toward our lit one, sometime in the year 2000. It only takes right about 3 years to travel the rest of the distance. In fact, the majority of its journey from halfway to here happens the last 9 months before it arrives. Zoom! It'll be cooking with its greatest speed by the time it passes. Once it crosses Pluto's orbit, it will only take about 90 days to pass right between Earth and the Sun.

17

X's inbound approach is being closely monitored by our best telescopic equipment on and off earth, but now small private observatories have started spotting it, which are not controlled. The most accurate calculation for X's next passage is now Springtime, or May, 2003. Ouch! I know. I wish it were not so soon also.

Many people have been lulled into complacency because the Y2K problems and a simple planetary alignment were over hyped; then nothing happened. Let me try to jump off the page here and slap you upside the face. It's the under hyped, covered up, and intentionally confusing one that is going to get you. The preprogrammed laugh track given to you by the media controllers many of you have been playing back and forth to each other every time the mention of a tenth planet or planet X comes up, is going to break by the end of this book if it hasn't already. This is it! Planet X is The One. It is going to Rock Your World. It's the Big Kahuna. The Main Event. The Real Deal. Let's get ready to get Rumbled! There will be no cable TV required or $40 needed to view this event. It will be live, life and death, in your face action. You will see it wherever you are whether you want to or not. God help ya. It's time to start praying....

Ancient History Revisited

Each time X approaches, it starts a chain of events that culminates with its passing by, causing our earth's surface to change abruptly. These changes are the cause of massive death and destruction. History is rich with these stories. Ruins of past great civilizations stand as monuments to these past events. The legends of Atlantis, Lumeria, Mayans and several others going under the sea or abandoned can be completely explained by the regular passing of Planet X. Just off the coastlines of many lands exist remnants of great sophisticated societies, such as found off Florida, Japan, the Mediterranean and more.

Many from the past didn't draw the connection between what looked to be a large comet overhead and what they were experiencing at the time (volcanoes erupting, earthquakes, land masses sinking and rising, tidal waves, severe weather, fires, floods, droughts and the accompanying food shortages). Some ancient writers simply noted what they thought was a foreboding sign or messenger of the prophets, when they viewed X in the sky instead of the cause for their woes.

Non-Religious 7 years of Tribulation

In 1995, Planet X got close enough that its far reaching magnetic and gravitational arms of influence started to affect Earth again significantly, which increases now daily. The closer it gets to us, the faster it moves toward earth, and the stronger its magnetic and gravitational effects become. There is an ebb and flow to these events that are akin to the calm waves before the next big set. Surfers would be most familiar with this analogy. Expect things to get worse as it approaches nearer but not in a straight line.

Our own sun's gravity is what is pulling it closer and faster along with X's own gravitational pull to the sun. Planet X never collides with earth or the sun because of the newly uncovered repulsion force which comes into play as they near each other. There is still great reason to be forewarned and concerned.

Earth now has two strong magnetic/gravitational influences in its vicinity (our sun being the main one), and its largely iron magnetic core is heating up because of it. It's like a car with the accelerator and brake on at the same time. Our sun is pushing and pulling on earth one way and Planet X is now upsetting this with its own set of influences. A delicate ecological, environmental and planetary balance between our sun and earth is increasingly being disrupted by the approach of X as it draws closer with each passing day. All the worldwide unusual weather that has broken all previous records and the seismic activity as of late is perfectly clear.

There's no need to waste time or millions of more dollars on equipment. Inquiring into the why or when things will happen by setting up and programming supercomputers are useless and futile. Today's melting and breaking up of massive glaciers and ice shelves, storms, floods, tsunamis, volcanic and seismic activity, fires, meteor showers, droughts and related energy shortages, etc. (you're paying for it now baby!), with all the loss of life, are only a preview of what is to come. Surviving massive death and destruction will make you feel more alive! You've been forewarned. You can do it.

The records of this approximate 7 years of turmoil before each regular passing are part of biblical record from at least the last three times it sailed by Earth. The Jewish exodus and Noah's great flood would be the last and third to last passing of X. With ancient historical, archeological science, weather, seismic data, and a small observatory, you don't need prophecy to predict the same cycle of events are happening again now during X's current approach.

You'd be clueless to ignore it! This is but a minor taste of what the passage will be like. And yet even these relatively small events are the cause of many fatalities and much destruction. Several areas of the world will begin to be severely affected as the date of passing inches nearer, so don't think you can wait to the last minute and then jet away safely.

Time is short. Now, in May, 2001, X is 2 years away and approaching from Orion. Check it out for yourself man! Find an Observatory and get someone to help you. Go and do a careful analysis of the world's weather and seismic activity over the last 5 years and see for yourself how dramatically things have changed from before that time. Then think about the media's explanation of global heating and cooling or sun cycles. Understand that one volcano can put up more hydrocarbons with one eruption than our world-wide dirty civilization can in a year. So we are the cause of only a minor part of the changes Earth is experiencing, and certainly have nothing to do with the up-tick in the seismic and volcanic activity over the last few years.

Denial Happens and It Reeks

The majority are out of the loop of knowledge and, even if paying attention, have been listening to a myriad of lies. I know it's more comfortable to turn away from an extremely upsetting reality or situation. The disinformation and awakening to Planet X will increase in direct proportion to its closeness.

First it was boldly, proudly and officially announced December 31st, 1983 that X was in fact finally seen! Then the media controller manipulators got wind of it and, as if by magic, the whole thing was denied, and suddenly a mistake or retraction had been made. Part of what they then said was that it was something else and that they really didn't see it. This sounded like, "I never had sex with that woman." Next, when amateur astronomers can absolutely see that it's inbound (which is happening now from small observatories) the disinformation will not sound the alarm, but will falsely reassure the masses with continued nonsense. At the same time, there will be hints given out to be wary of something for people with an ear to hear them. Arguments among scientists may attempt to lull the public to sleep and warn us at the same time. Both tactics, more than likely will be used to assure that panic is kept at bay as long as possible.

The situation is too dire to reveal it publicly straight out by the scientists, leaders and government officials in the know that want to. They all fully understand what it would do to our society if they came clean with the facts over the major media. A complete global financial meltdown would ensue immediately. What would you do right now if your cash and bank balances were suddenly worthless? Can you imagine the amount of death and disruption just the announcement alone would cause? Pay attention here: The story will not be told or given out accurately, no matter how much you believe, hope or imagine that it should. Your "they will inform and take care of us" logic is out the window on this one. It's the ultimate Catch 22! These people are at the top of the heap, in the know, and just sitting on the information. It smolders underneath their feet with every step they take. I'm sure some are driving themselves a little crazy with guilt. Others may be hitting the bottle. How would you feel in their positions? I wouldn't always be too quick to judge them. Now watch as I do just that.

The financial controller gangsters with the best information money can buy are at the top of the heap. They are the ones paying for the lies and promoting the

20

grass- roots awakening, while watching the approach and its effect on earth carefully. If you're beginning to become frightened, that is a good thing! Now take control of that emotional energy and use it constructively to formulate a plan to save your life. The supermarkets, gas stations, electric companies and telephone service will be a thing of the past for years afterwards in most areas.

First the effects of Planet X were said to be caused by global warming. As predicted, when too many scientists world-wide challenged this, the blame or story was shifted somewhat to super sun cycles, even though global warming is still being used as a reason. Sun cycles have never been the cause of simultaneous weather and seismic activity. This big lie theoretically states SUPER sun cycles are the cause. Of course, it involves a period of time so long, there's no possible way to verify it. Our sun isn't causing Earth's problems, nor are we for the most part. It is the increasingly strong magnetic and gravitational relationship between Earth and Planet X that is.

Now here is where some of you will say that the magnetism between planets is small compared to their relative size. The fact is; the 30 mile thick crust of earth diffuses the magnetic readings on the surface. This hides the true relationship between the heavenly bodies. Planet X would not be heating up our core again and causing so many changes, though billions of miles away, if the magnetic relationships didn't truly reflect the planets' sizes. I wouldn't be so overly proud about our so-called modern-day sciences being correct on some heavenly issues. I strongly suspect that ancient earth people may have had much of this figured out more correctly than we have today. You'll just have to use your own judgment on this and step away from what the so-called authorities of modern-day scientific understandings are teaching on planetary magnetic relationships. A willing suspension of belief is in order on this item at this juncture.

HELLO, HELLO, SATELLITE CELL PHONE NOT WORKING!

Watch what happens to all of Earth's satellites a few months prior to X's passage. Can you say, "Hello? Hello? My cell phone, satellite TV, and landline-based satellite long distance are not working!" Then think for yourself. If X's gravitational / magnetic influence will render useless almost all of Earth's satellites <u>during the months prior to passage</u>, what will it do to Earth when it passes? <u>If you read and forget this entire work but remember this one piece of information you'll know it's time to migrate inland and upward if you haven't already</u>. Imagine it's a few short months before Spring 2003 and your cell phone is all static. Are you going to listen to the government's talking heads' red-faced lies and bullshit, putting your life on the line, believing the nonsense dished in your face yet again? The governments will be full of excuses, lies and nonsense trying to explain what's happening with the satellites. Our sun will have nothing to do with this disruption. That can easily be checked by trying to use your cell phone at night and seeing if the interference is still happening. If it is still on the blink, you know it has nothing to do with our sun. They know they will never be held accountable for anything they say, so they'll say anything they want. HELLO HELLO? Are you listening? Or can't you hear a thing.

Understand what we know, what we think we know, and what is true are constantly changing in the scientific community as our knowledge increases. There are many theories that are regarded as fact. <u>The fact that we can do math correctly to figure out where a planet is going to be in time and space does not mean we know the mechanics behind why it's moving.</u> The why for planetary relationships is theory, and not proven, while the math may be only just true enough. Since we have figured out the orbital math, we seem to be smug about the theories that truly only speculate on the how and why planets move and what their true relationships are.

The granddaddy of all conspiracies is this: Using our tax money, the controllers are doing their best to hide the news of X's approach and increased earth disruptions, as they ready themselves for the inevitable poleshift. They give us a series of smoke and mirrors in return for the money they take from us. Do you really think that you'll be invited into their shelters when the time arrives? Can you imagine where a large portion of the billions in just Star Wars defense money is being siphoned off to and for?

There are factions within this group of insiders that would like this story to surface. These concerned people are only allowed to create false alarms and give information out via obscure forums to further public awareness. The majority of the population will be uninformed. Most will be in unsafe areas when the catastrophe takes place.

Occasionally leaders who are in the know and told to stay quiet will speak up when they are not supposed to because of guilt. Their consciences just bother them so long that they let the cat partially out of the bag publicly while not entirely planning to do so. Now if someone pulled a cat partially out of a bag, you'd still know it was a cat, wouldn't you, if you were looking? This will happen more often as the time approaches, so keep an eye out for it. Unless you know what to look for, these slip-ups will go unnoticed by the vast majority. The following excerpted press release came over the Reuters wire service. The media handlers determined it was too hot to handle, that to try to cover up this evidence through disinformation, would be much more revealing than to say and print nothing more about it. Sometimes news gets out that is so pointed, accurate & revealing that the only way to handle it or to cover it up is with stone cold silence. The major media handlers were not allowed to do any follow up stories, so the following announcement faded away from view. It was hidden among so much other information put out day after day that the silent treatment worked perfectly. The disasters and consequent massive population shrinkage that the Russian leaders are talking about will affect the entire world, but they don't mention the rest of the world, because that is not their concern.

Russia Sets Out to Tackle "2003 Problem"

By Andrei Shukshin, *Reuters*, Sept. 13, 2000:
Russia's Parliamentary leaders and President Vladimir Putin agreed Wednesday to embark on a three-year crash course to thwart what they said was an <u>anticipated chain of disasters due to hit the country in 2003.</u> "(These are) issues of extraordinary importance, strategic issues which may degenerate into <u>a serious threat for the existence,</u> — I want to stress this — <u>for the existence of Russia,</u>" former Prime Minister Yevgeny Primakov told reporters.

   Pro-Kremlin Party Brought Up The 2003 Problem: Boris Gryzlov, leader of the pro-Kremlin Unity faction which was the first to raise the issue, said Russia would also have to deal in 2003 with a <u>massive population shrinkage</u>. Gryzlov said the problems had already been discussed with cabinet ministers, and the parliamentarians had agreed with Putin to set up a commission to tackle the issue head-on. "The question was discussed at length, and the president approved our initiative and said he would dispatch representatives of his administration to the working group," Gryzlov said after the Kremlin meeting.

   A couple of things to note in the above wire are that the Russian leaders were worried that these disasters and massive population shrinkage might be so severe that Russia might not exist as a country afterwards. I am astounded that they were so flagrant and used the phrase "massive population shrinkage." The new non-communist Russia may be less controlled in some ways than the United States. They are saying here that <u>they expect a huge part of the population to die in 2003.</u> What do you think from? How would they anticipate exactly 3 years ahead of time without some sort of inside information they are withholding? The writers for the media who are out of the loop or told to fill in the holes of this announcement came up with 3 possibilities: aging population and infrastructure, plus a large debt due that year. None of these 3 possibilities covered up or saved this story from being what it is, "truly an eyeopener." Now let me think for you here about these 3 conclusions. How could you predict 3 years ahead that suddenly in 2003 elder people were going to die at such a massive rate that Russia might not exist as a country? This is completely ridiculous and an obvious whitewash. Would you also be able to predict that an aging infrastructure was abruptly going to deteriorate in 2003, such that Russia would not exist afterwards also? — not unless you knew absolutely the time frame of an impending calamity and its cause. What difference would a large debt due in 2003 make? Why would this matter? Banks renegotiate debt for foreign countries constantly. Having inside knowledge of the pending events, the big bankers are calling in most all their debts in 2003. This plainly isn't going to cause the "massive population shrinkage"; it's because of its expectation.

The Russian cabinet ministers and parliamentarians agreed with Putin, set up a commission and are tackling the issue head-on. What are you doing to prepare? You've got to <u>read between the lines and have a little inside knowledge, of which I give you enough in this short book</u> to know what's being given out publicly here. Wake up, man! This is real. I didn't pull this article out of my hat.

Please do your own research with a wide eye open toward the continuing disinformation. Some of the same sources and information I offer here will be used mixed with rubbish. The simple clear truth will be covered up by mixing lies with truth. Be careful if your mind is just hoping it will all go away. Hope exists in your ability to bring yourself up to a state of mental certainty and use clarity to help yourself and loved ones survive. Do your own research for this reason. Find your own unique sources that only you can. Then get this story out the best way you know how to save lives. Look inside and trust in your heart for the truth on this issue. Also, please give away this book as soon as you're finished. It does nobody any good lying around.

My Story

I do not hold any degrees or weight in the scientific community. Geez, ain't that a relief for me with this hot potato? I'm glad to be a high school drop out revealing this situation. When I attended college for a couple years in my late 30's, the last semester I received straight A's and got on the president's honor roll. I'm glad I dropped out right after that before graduating. I found out after my first computer programming class, I didn't want to do that all day long, day after day. Now, for all those who want to attack my credentials, I'm pleased to say I've none to go after.

The controllers' strategy is to get the story out through obscure sources. I fit right in there, being simply the son of a musician. I was raised in Los Angeles in a community of some influence to the entertainment industry. The section of the San Fernando Valley called Toluca Lake was my home. Andy Griffith lived next door and Bob Hope down the street.

I add pop media phrases and lingo throughout this work. The media-controlled generation is in part whom I'm trying to get this message out to. These younger people have the strength to adapt in the aftertimes.

My father's a movie and music producer, singer, songwriter, actor, author, director and all around cool guy named Lee Hazlewood. He's worked with (among others) Frank and Nancy Sinatra (mainly Nancy). His most notable material was a #1 hit song he wrote and produced and then was performed by Nancy titled "These Boots Are Made For Walking." ("One of these days these boots are going to walk all over you," is a line from the song.)

There were several other hits Nancy and Lee had together. Lee also produced a beautiful tune with Frank that was titled "Something Stupid," and it hit # 1 on the charts too.

I had every opportunity to get into the entertainment field. The most I ever did was practice singing and acting and write a few poems. Given the success I grew up around, I felt that I could do anything I wanted, so I naively went about trying to make this world a better place.

Back in the middle 1970's, at the age of 19, I started work with a budding solar corporation. It had the technology to create photovoltaic solar cells with efficiencies that could undercut the cost of fossil and nuclear fuels in the creation of electricity. After a dozen years of working on and off with this company (because of financial difficulties), it became clear that we were held in our tracks by the controllers. They effectively used several different legal and financial maneuvers to stop our efforts.

There still exists a way to create electricity through photovoltaics, which is under the price of fossil and nuclear fuels, although now I'm sure even this technology is outdated compared to what has been developed and withheld from the energy marketplace. There also exists a way to cheaply convert any vehicle to run off of liquid hydride (safe hydrogen). This can be inexpensively created with this solar technology. It all worked together to form a complete 24 hr. alternative to the madness and pollution of oil and nuclear power.

This holding back and shelving of cleaner, safer and less expensive technology for personal interests is all in a day's work for the controllers that have their vested interests in mind. Raising the standard of living or quality of life and wiping out pollution around the globe was the primary goal of the company I worked with. This would have taken a vast amount of wealth and power out of the controllers' pockets had our company succeeded. Now the 100's of articles about this company rest in dated publications of The Wall Street Journal, Washington post, New York Times, Los Angeles Times among dozens of others. It's old news and at present not very important at that.

The Twilight Zone

It is now past the year 2000. You may be aware that disinformation is a part of the landscape of this world. Some very real issues have consistently been swept under the table as part of this campaign. I'm going to introduce you to a few that are perhaps unfamiliar to you. I urge you to keep an open mind if this happens to be your first exposure. This world has entered into The Twilight Zone with only some people realizing it. In reality it's been there all along.

Developing your own insight and intuition is now imperative for your survival. The ability to know what you should do from moment to moment in the middle of lies and chaos has to be developed by each individual. Each person has to find their own unique way. Turning inward to find answers to questions is a skill that you need to begin doing or attempting to do immediately. Some of the subject matter that I'm going to bring up in the past you might have ignored. Take everything presented with an open mind. Understand your mind has been manipulated without your being aware for purposes of control. Certain subjects have been taken off your table because these controllers believed that they could exploit you better without it. You've been taught through media manipulation to scoff or disbelieve what is common knowledge elsewhere.

Finding and recognizing your direct connection to a universal collective mind that exists or at least your inner self needs to be turned on high from this point on in your life. There is very little time left. Read this book with much more thought and depth than you would the daily news. Getting used to the unusual signposts or subject matter in the Twilight Zone of this world is the only way you're going to survive if you plan on being a part of the next — and there will be a next. The regularity of Planet X's passage and destruction has not stopped the world from reaching now 6 billion plus people. So don't take a defeatist attitude and say it's not survivable. It may in fact not be survivable in certain areas, and that is all you can conclude. Most of the society you've grown accustomed to is coming to a quick, violent end in 2003. Some places sooner than that will find disaster at their front door.

I cannot urge you enough to start finding your second sight now. This is a unique task and is different from individual to individual, so I can't explain how to do it or expect my way to work for you. Perchance only letting you know it is possible and then you creating the DESIRE to do so is all it will take. After finding this inner voice or second sight, your whole view of the world will change. It's the difference between just owning a computer and having one connected to the internet. With your own inner voice, the challenges of the next world will be overcome, and the opportunities, excitement and adventure will be immense. Again, from this point on, you need to set aside any disbelief until you've finished with this book. At that time ask, find, pray for, or question your higher, deeper, or inner self and see if you can get an answer in your own unique way. What your conscious rejects will not always affect your unconscious. This part of your mind will begin to work, figure things out and put it together for you, whether you realize or not. This could all happen in time enough for you to bust out and away for your survival.

Controllers of old in San Francisco

The great San Francisco Earthquake was first known to have happened in 1868. The Hayward fault moved and destroyed San Francisco. The government controllers of the day commissioned a study to find out why it had happened to see if it could happen again. The geologist found The San Andreas Fault to the west and the Hayward Fault to the east of San Francisco. The conclusion was that the relatively small city should not be rebuilt or encouraged to grow, being between these two major faults. The powerful financial interests (controllers of the day) destroyed this report. This conspiracy led to the next great San Francisco quake which most are familiar with in 1906. The conspiracy continued after this second great San Francisco quake. The newspapers of the day said that only about 500 people had died when, if fact, more than six times that perished. They also decided to call it The Great San Francisco Fire to play down the faulty reason for it. Real estate prices and future commerce are the reasons for the cover-up as usual.

This is a minor conspiracy compared to what we face today. Our whole planet will be again changed dramatically by the passing of Planet X. Again it's the powerful financial controllers of today that want to keep what they have going as long as they can before it all collapses at the expense of billions of lives. The cataclysmic earth changes caused by X passing will change the world so completely; these controllers know they won't be held accountable.

1985 Mexico City Earthquake

There are documentaries that explain this now openly on TV. Once enough time passes, the cover-up gets out with nobody caring about it anymore. There's no further need to withhold the information. What people don't seem to care about or realize is this same process of media manipulation is ongoing. The process of covering up did not stop because you learned about it happening in a past event. The 1985 Mexico City quake official death toll was 6000. The real death toll was well over 10,000. Again the reason was big financial interests (controllers) wanting to play down the reality for future commerce, such as tourism, and to hold up real estate prices. These are just a couple of the many past cover-ups that are common knowledge. This does happen on a regular basis for reasons that can only be described in one word "greed."

Florida's 1992 Hurricane Andrew

Find and talk to the survivors and individuals who were directly involved in the checking of wreckage for bodies. Any persistent researcher can find the truth of out of the thousands lost in this disaster.

First be aware there were officials that were down at Homestead while the hurricane was going through that knew zero about what was going on right in front of their noses afterwards. Calling up an official or two will not necessarily give you the truth. You may get their limited view only.

If you choose to research this story, make sure you ask enough people pointed questions that you interview. You will absolutely find out that thousands died. When questioning, probe deeply. Did they just drive through, or did they get a chance to walk through the neighborhoods before the cleanup crews disposed of the bodies.

Contact ambulance drivers. Many saw several of the refrigerator trailers stacked-up with hundreds of the dead bodies in them. Some honorable National Guardsmen had direct knowledge of the cleanup and can give you the real story. Thousands of lives were lost. In one apartment building alone, only 12 people survived out of 810 units!

The best weather reports had Andrew hitting north of Homestead, Florida until 11 PM. One hour later at midnight the winds started. Since there were no mandatory evacuations for this area of 400,000 people, most decided to stay home and ride it out.

A wind speed indicator broke when it registered 214 miles an hour at the Homestead Air Force Base! There were 8600 of just trailer homes alone wiped off the face of Homestead, Florida that night. Again the majority of these families were home at the time. Do a little math. Now think about all the additional flattened neighborhoods, 1000's of individual homes, the apartments plus hotels where the rest of the 400,000 were at. Many stayed and died.

For weeks afterwards much of the effort going on was cordoning off the area, sealing it up and keeping it from view. Their excuse of course was they didn't want to allow looters in. This was solely an affront agenda. Each day hundreds of bodies and body parts were found, picked up and loaded into closed refrigerator trucks. This process proceeded over and over again, day after day for several weeks. Helping survivors was not part of the program of the cleanup thugs. In some cases, weeks went by with no help to the people who needed it most.

28

What the media was allowed to show through the news outlets about this storm tragedy was completely sugar-coated. Truckloads of refrigerated bodies and body parts were carried off to makeshift crematoriums located at the garbage dumps. If you are the slightest bit skeptical, I challenge you to make a trip to Homestead Florida and start questioning people who survived that are still there. There this knowledge is common and the stories endless. It may be worth it to you to find out without a shadow of doubt how the controllers manipulate the media. Much of the effort was to hide the large death toll. Hundreds of badly hurt people died because getting help to them was not the priority. Protecting the financial interests of Florida was all the cleanup crew leaders had in mind. After swallowing this hardball, you will have a clear idea about how the controllers are handling the knowledge of the regular passage in 2003 of Planet X and the expected wipeout of the majority of this society.

If you ran across the so-called authorities that were in charge of regulating this clean-up or cover-up work in South Florida, they were dressed in plain black with simple badges on. If you seemed in dire need of medical attention and asked for help from them, these minor authority figures might have put a gun to the back of your head. Next, they would tell you to shut up and mark your social security number on your person. They would then explain to you, next time they came around this social security number would be needed to identify your body; when they pick it up! This was all the help one could hope for from these guys. These paid-for goons had no time to help survivors in need, because the work of cleaning up and covering up bodies was their only job.

If the true extent of this disaster got out, the property values and the tourism market would have dried up. Florida's controlled financial interests couldn't have that.

Some of the in-the-know National Guard that helped out had a total body count of 5200 circulating amongst themselves. What our controlled media is still saying is anywhere from 22-57 dead. Wake up, world! Did you see the devastation from the helicopters of neighborhood after neighborhood completely destroyed? Remember, many of these people were home at the time, expecting a storm to hit north of them. Again there were no mandatory evacuations in this area. On top of this, most were complacent, having never seen a killer storm like this in their lifetime experience and simply did not know what to expect. The insurance companies didn't even have to pay 100s of families that were completely killed. When nobody is left, they say hooray, no one to pay.

Who Cares?

I mentioned the Mexico City and SF events because they are very well known. The deception is now documented and being presented from normal broadcast and cable television as, in fact, what it was, cover-ups or conspiracies. Andrew was fairly recent. Who cares? — you might say, to keep the money rolling in, a few thousand deaths were covered up at different times and places. They didn't actually do any of the killing. Mother nature did. It does make since to save jobs and money.

In addition, hundreds died after Hurricane Andrew, because no help was forthcoming. Gathering up other bodies and body parts was the main priority. Withholding the information about Planet X arriving is not about just a few hundred or thousand people. We're talking about billions of lives. One of the ways for you to understand that this these cover-ups are ongoing is to view the history of how similar, smaller events have been handled previously.

It should be clear to you, that it's all about money and power. The people who control the international corporations run this world and not the puppet governments. MONEY IS FIRST AND PEOPLE'S LIVES ARE SECOND. That's just the way this world works.

The information about Planet X is looked at by the people who run the international corporations in terms of their bottom lines. For the most part if they don't keep X hidden from most, they lose too much money now and to hell with later. Since X is coming, making money later is questionable. All bets are off. Saving lives is not tops on their priority list. And that's as real as it gets.

I do understand that there's a large percentage of earth's population that will perish no matter what is said or done. In spite of that, there are many millions of lives that could be saved if the powers that be decided to open up the information superhighway and media machine to Planet X.

Gangsters — They Are Rulers Of Our World

There are many names given to these secret groups of people who have a significant influence over our thinking, money and lives. I could give a list of companies, individuals and family names that I'm sure would ring a few bells. I choose not to repeat these names that you may have heard or read previously time after time. The names and labels of their semi-secret little organizations and families are irrelevant. Some of them are the same names of approximately 200 that control a private trust in Puerto Rico that all your IRS tax money goes into.

Most people have no clue that the IRS is a private company along with the Federal Reserve Bank that this trust pays into. The only benefit that the US government gets are the "for profit bank loans" from this private institution. Income Tax is a voluntary system, and you give up your rights when you sign a 1040. So, not only are we citizens being coerced into voluntarily paying a private company, we pay them again for interest on those loans. What a racket! It's a scheme to defraud, and they have the majority of the US court system in their pocket to enforce the scam. Actually people win against the IRS all the time. Part of the deal is always to shut up. I caution you to do your research first if you want to challenge paying income tax. If you don't know how to handle them, these dogs can bite. When the IRS goes after your property and assets, they do so illegally without a court order or proof of wrongdoing. Many of you will be sickened to find out this is true, and still others just won't believe it. The media monsters have just been way too effective in perpetrating this lie. Many individuals have sued the IRS when finding this out and have been paid back the taxes they paid to the IRS with interest.

The income tax system is a passé issue compared to what we're looking at right around the corner. The IRS will be gone soon, but so will most of everything else. I only bring up my knowledge of the IRS so the few that do know this will understand that I've done my homework. I understand how this system we live in works, who's on top, how they play the game, and, most importantly, why most will not hear of the approach of X until they see it with their own eyes. These controllers don't want the majority to know. You should feel like the privileged minority if you find out, comprehend, and accept this truth in time enough to save yourself.

I've got a buddy of mine named Jay. He comes from a well-to-do family. They're all highly educated and articulate. Jay's father is a rather wealthy retired executive. Jay is very bright, exceedingly educated and even a creative musician in his 30's. He is one of the few people who knew about X before I met him. For Jay it cannot arrive quickly enough. Jay and his father dislike some of the people and institutions of this society so much that they'd be a lot happier if X arrived tomorrow rather than 2003. I've got to laugh with him sometimes about his attitude, because there's a side to me that most definitely agrees with him. Truly, what is wrong with this society simply can't be fixed or improved. It has to start anew. Besides, who am I to question God's plan? I might as well just find the positives in it.

The chances that you will run across any of these controllers are slim to none. Like I've explained, I prefer to identify them by what they try their best to do, control. Power is their reason to live. Some of these controllers have the same mindsets as street hoods. They feel first of all that power is something that they can't get enough of. No matter how much money or influence they have, they are never satisfied. In order to get more of these two, they are willing to lie, cheat, steal, murder and destroy the environment. The world is not enough for them.

These people look like legitimate businessmen. Many have families that are at times unaware of what cheap and uncaring, street-gangster mentality these people hide from them. They possess the right business suits, image, lingo, mannerisms, untold number of expensive adult toys, club memberships, connections and some with more money than they could ever spend. The reserves they govern are enormous and carefully hidden. This wealth does not cover up their street gutter way of thinking if you look close. They are never happy with any monetary quantity or level of power. They feel, because of their worth, that they are above the law and can get away with and step over anybody. I apologize to some street gangsters for my comparison. Some of these people are a lot more ethical and down to earth than these corporate controller gangsters.

Next time you see a conservative-looking man in a business suit, don't assume anything about him. He might very well be another thug blending in to the corporate world, doing dirty deeds behind closed doors. These egomaniacs are interconnected within giant international corporations across the globe. Some are also part of various secret, loose-net organizations. Others are in politics, royalty, law, entertainment, media, banking, pharmaceuticals, illegal drugs, energy, NASA, military, the so-called justice system and many more. It stands to reason anywhere you find big money or power you'll find them. They all don't necessarily know each other. The fact is clear that certain people do their best to control whomever and whatever they can for their own selfish interests.

There are also some that hold no positions or titles and may in fact be the most influential and powerful of all. These people may have greater power in their position of anonymity. With their vast hidden wealth, they spend part of their days putting people in positions of power. These people have the titles and status in all and more of the above-mentioned areas but not the true power. The controllers play them like puppets on so many different strings for great financial and ego rewards.

The controllers know that the world is black, white and gray. There are straight arrows that resist being controlled in the world's power game. I guess we can give thanks to a few that do look out for others. Still some will play the puppet game only partially. Then, of course, there are many who are in it up to their necks. These are the ones that are fully aware that not to play the game would mean their necks.

Many controllers are at times as ruthless as you can imagine. Cover-ups, threats, blackmail and murder are just part of the game. If it comes down to murder, making it look like accidents, suicides or natural causes are part of the skills of the people that are hired. Sometimes to make a point, they make sure it looks like murder. Being at least one step removed from this contract about who and how it's supposed to be done keeps their hands from ever getting dirty. Covering up the tracks of the ones on the front lines doing the dirty work is done by their media puppets. This keeps certain facts about the death out of the public's eye.

This is how the control game is played at its worst. Most of the time there are easier solutions when initial cooperation is not forthcoming. Direct or implied threats work just as well the majority of the time. Even just the threat of future lost or lowered income is enough to keep most of the people at the ends of the puppet strings playing the controllers' money and power games. Media are the easiest. Tell the wrong story once too often and you risk losing your license or more.

Sometimes a fall guy is created and he knows full well his role ahead of time and what it would mean if he didn't go along with the program. If you're aware or well read enough, you already comprehend how things work in our world. There are some who turn away when presented with harsh realities mumbling "conspiracy theorists." Well, I challenge you to find out for yourself the truth of these matters. If you have your head buried deep in the boob tube or are just getting by from week to week with a job and family, I can understand why you think the world is Father Knows Best and Mr. Rogers Neighborhood with aliens being a silly idea. Even some well read and much too busy intellectuals have missed the boat on these issues. It really only takes paying a bit of attention to see at least a hint of some of these characters and situations being played out in this society.

Today the most influential controllers keep the live feeds to the Hubble telescope and Soho satellite from us for good reason. Planet X is almost here. They have already paid others to prepare themselves for their future survival. These people are not telling the public what they know. They even deal with each other on a need to know basis. They do have an enormously difficult Catch 22 to deal with.

If I were these controllers, I would do the honorable thing to save lives. I would announce the impending disaster today. Again this would ruin the financial and real estate markets instantly and start a depression world wide. Without money or credit there would be much death, starvation, panic and civil unrest well before the actual event. For people like myself that value human life, an honest bulletin of the pending events would be well worth it. As long as the airing led to saving more lives than the amount of death honesty would bring, then it would be welcome. A major immediate effect would be to get a large percentage of the population out of harms way and relying on themselves to survive well ahead of the shift.

Imagine if an honest announcement was made. "Well folks, the world as you know it is about to be largely devastated by a shift of the poles caused by the strong magnetic or gravitational pull from a brown dwarf star that's scheduled to pass by in the spring or May of 2003 on it's regular orbit. Sorry but we don't possess the resources to help you. You're on your own. We've known about it for a couple decades but have withheld the information from you for the purpose of maintaining society's structure, which translates into greed and control purposes."

I suspect a few people would be a little irate at this late disclosure. This is one of the minor reasons for the cover-up of the most important news in all history. Tipping their hats risks the majority of everything they possess and control, which is of course slightly closer to their main concerns.

These people are already prepared for the aftertimes, so why would they want to ruin the world they control now while it's still in their hands. They don't want people finding out where they have set themselves up or give them time to search them out. The very last thing they want is for droves of hungry refugees on foot with no gas for their SUV's ending up outside of their well-hidden enclaves. Maybe if they cared about the millions of lives that could be saved, they would throw caution to the wind and come straight out with what they know and let the chips fall where they may. NOT! It ain't gunna happen. If they had any heart or courage, this would most certainly be the honorable thing to do. Heart, courage, and honor, where do we find these traits in this world of money and power today? Too few places is the broad answer.

This story has already gotten out from several different sources, and the unusual twist is the controllers want it to! Inside these groups of insiders in-the-know, exist a few that do care some. The greediest wield the most power, however, so the truth is simply leaked unofficially in several ways. These leaks are many — otherwise I wouldn't have run across this information myself. Do you really think the most important story in the history of mankind, that will touch the entire world's population, could possibly be covered up completely? My having discovered this is not all that special. What may be somewhat unique is that, after I stumbled onto the story and checked it all out from every angle, I then decided to present the information back out to the public at large. I did this in a way that adds a lot of pointed conciseness, clarity and even a little entertainment value to it. Even the end or new beginning has to be sold if you want a chance to make a difference. In this case a self-published, underground, and hopefully popular book would result in saved lives, and that's what this here is all about. The kids have to be given a chance to create their new world. Most aren't old enough to know much about this one yet. Good thing huh? I don't know why, I just reread the last 3 lines I wrote and burst out crying.

These controllers are using one of their most sophisticated tactics to confuse and warn the public at the same time. The smoke and mirrors tactics are part of how they're hiding and revealing it. If a story is so big, shocking and has so many ramifications that it simply can't be told truthfully, letting it leak out and shading it with half-truths is the way it must be presented. This also has the added benefit of making sure that nobody gets held accountable for the cover-up. Partially false stories tainted with a heavy dose of skepticism are created related to what they're covering up that contain truth, half truths, and lies. This effectively confuses the reading public as to which one is real. When you do hear the truth, many tend to remember where you heard it before. Let's say the first time you heard about something, the writer tended to make it sound implausible, embarrassing or silly. More than likely if you do stumble on the pure truth of the matter, that previous memory still taints the correct facts that are right in front of your nose.

How many people read? How many people read this sort of material? The controllers know and want this story to get out to a certain extent only. They do control the major media in a large way. This gives them the ability to simply drown out or rewrite and make questionable any news story that comes too close to the truth. **With all their available media outlets, they create news if need be to divert your focus. Major news dramas are now being created to keep your attention away from Planet X!**

We are a culture of fast food and headline news. Deep stories are missed by the masses. They're just too busy working, paying bills, eating, sleeping and trying to find a few minutes to lose themselves with whatever sort of R & R that they're into. Even if people are capable of understanding and hear it, the attitude of living day to day prevails. Honest announcements on TV and in the newspapers' day after day would still leave a large percentage of the population ambiguous. Many simply would still not listen. Only when it's staring them in the face, might a move be in order. At that time they might become one of the many that may miss being killed in the initial events but die in the aftertime because of lack of preparation.

Truth, Smoke, Mirrors, Half-Truths, and Lies

The following are three reprinted articles reporting the same information. My comments about these articles will appear before and after each one. The scientific study for article # 1 printed below was generated by The Open University in the UK. As you will see, it was in need of covering up because it gets too close to the heart of the matter. Also, it comes from the scientific community, which is a major reason for trashing it. A musician's son can say whatever he wants, and who the hell cares, but the gods of science must be snubbed. This information was found by studying comet trajectories. This makes it another new, unique and completely independent source for further evidence of Planet X.

Massive planet lies beyond Pluto
ROYAL ASTRONOMICAL SOCIETY NEWS RELEASE

The mystery object could be bigger than Jupiter, our Solar System's largest known planet. Intrigued by the fact that long-period comets observed from Earth seem to follow orbits that are not randomly oriented in space, a scientist at the Open University in the UK is arguing that these comets could be influenced by the gravity of a large undiscovered object in orbit around the Sun.

Writing in the issue of the *Monthly Notices of the Royal Astronomical Society* published on 11th October, Dr. John Murray sets out a case for an object orbiting the Sun 32,000 times farther away than Earth. It would, however, be extremely faint and slow-moving, and so would have escaped detection by present and previous searches for distant planets.

Long-period comets are believed to originate in a vast "reservoir" of potential comets, known as the Oort cloud, surrounding the solar system at distances between about 10,000 and 50,000 astronomical units from the Sun. (One astronomical unit is approximately the average distance between the Earth and the Sun.) They reach Earth's vicinity in the inner solar system when their usual, remote orbits are disturbed. Only when near to the Sun do these icy objects grow the coma and tails that give them the familiar form of a comet. Dr Murray notes that the comets reaching the inner solar system include a group coming from directions in space that are strung out along an arc across the sky. He argues that this could mark the wake of some large body moving through space in the outer part of the Oort cloud, giving gravitational kicks to comets as it goes.

The object would have to be at least as massive as Jupiter to create a gravitational disturbance large enough to give rise to the observed effect, but currently favored theories of how the solar system formed cannot easily explain the presence of a large planet so far from the Sun. If it were ten times more massive than Jupiter, it would be more akin to a brown dwarf (the coolest kind of stellar object) than a planet, brighter, and more likely to have been detected already. So Dr Murray says that such an object will be planetary in nature and will have been captured into its present orbit since the solar system formed.   (end of article)

He's right twice again here. It's a brown dwarf star, and it has been detected. He may be unaware of the 1983 sighting and the cover-up, but it's very telling that he's got this amount of correct information through a completely separate scientific venue.

Now the above article and study and another one from a completely different source were completely trashed by the disinformation wrecking crew. Much confirmation from the scientific community comes to the surface only to be drowned out by the media controllers with rewrites.

Article # 2 tries to present the facts in such a way as to make them seem implausible. It is a lot longer and was much more widely circulated than article #1. The reason for this was to dampen the impact. Article # 2 does a good job of making you ignore what's there, which is the entire reason for it. It is also completely tainted with words of speculative disinformation and much more widely published. Look closely at the words they put in all caps to see their motivation for writing the article.

The powers that be even got the inside scoop that article # 1 was going to be published, so article # 2 was published four days prior to article # 1. They mention a bizarre cult and use dissenting words and phrases like "hint, mystery, hypothetical, hypothesis, fringe phenomena, more questions, science fiction, spookier." It says that any object like this would take millions of years to orbit. This means they want you to ignore it in your lifetime. It even mentions that the IRAS satellite in 1983 should have seen it if it existed, but didn't. I guess they missed the announcement from the chief scientist of the IRAS Satellite that found X. The disinformation that followed must have been more compelling to them. They make it a point to mention that no telescope has ever seen the object. I guess that part of their disinformation is starting to unravel with the latest sighting from small observatories. A point is also made that this object is not a "Nemesis." Nemesis is the precise name the ancient Greeks gave for X.

The controllers' thinking is this: It's better to trash a nice piece of evidence from the scientific community than let it sit untouched for others to begin to ponder and worry too much about. Too many might see the truth too quickly and disrupt the status quo. Also what is the easiest way to catch someone in a lie or cover-up? They give too much information. Article #2 gives more information than the whole point of the study that article #1 tries to make. Comet storms weren't even mentioned in the original study of comet orbits; neither was the IRAS satellite. It also says speculation about another planet in our solar system dates back earlier to Pluto being found in 1929. Hummmm. The people that wrote article #2 try to make the study and conclusion from article #1 look like pure speculative nonsense. However, clearly they seem to know a lot more about the subject they're trying to suppress than simply what the scientists behind article #1 are concluding from their study.

Here it is. Trash or conspiracy journalism at its best, doing the dirty deed of covering up the real deal. Are they adding more information than they should have? Is this one way the controllers are getting X info out? Think for yourself, you know my opinion.

A Mystery Revolves Around the Sun

Researchers suggest that a huge unseen object orbits on fringe of solar system.

By Alan Boyle MSNBC, Oct. 7, 1999 — Two teams of researchers have proposed the existence of an unseen planet or a failed star circling the sun at a distance of more than 2 trillion miles, far beyond the orbits of the nine known planets. The theory, which seeks to explain patterns in comets' paths, has been put forward in research accepted for publication in two separate journals. SPECULATION ABOUT the existence of unseen celestial companions dates back far before the discovery of Pluto in 1929 — and even figures in more recent fringe phenom-

37

ena such as the 1997 "Heaven's Gate" tragedy and talk of a new "Planet X." This latest hypothesis, however, is aimed at answering nagging scientific questions about how particular types of comets make their way into the inner solar system. Some comets, like Haley's Comet, follow relatively short-period orbits — circling the sun in less than two hundred years. These comets are thought to originate in the Kuiper Belt, a disk of cosmic debris that lies beyond Neptune's orbit.

The best way to think of the distances involved is in terms of Astronomical Units. One AU is the distance from Earth to the sun (93 million miles or 149.6 million kilometers). Pluto, the most distant of the planets, is at 39 AU. The Kuiper Belt extends from 30 AU to perhaps 1,000 AU.

Even further out is the Oort Cloud, a spherical haze of comets surrounding the solar system at distances between 10,000 AU and more than 50,000 AU. That's where long-period comets such as Hale-Bopp are thought to come from. For some time, astronomers have noticed that the directional patterns of these comets are not completely random. And after years of study, some researchers are reporting that the patterns hint at something big out there perturbing the cometary paths. No telescope has yet detected this object. But on the basis of its gravitational effect, John B. Murray, a planetary scientist at Britain's Open University, speculates that the object could be a planet larger than Jupiter, the biggest of the solar system's known planets. Murray puts the object's orbit at 32,000 AU, or 2.98 trillion miles from the sun. His proposal appears in the Oct. 11 issue of the *Monthly Notices of the Royal Astronomical Society*.

Meanwhile, researchers at the University of Louisiana at Lafayette say the object could be a planet or brown dwarf — that is, a dark, failed star — roughly three times the size of Jupiter and orbiting at 25,000 AU. The researchers, led by physicist John Matese, say their paper is to be published by the journal *Icarus*. Both studies acknowledge that other factors could influence the pattern seen in long-period comets: for example, the Milky Way's gravitational tidal effects. But the Louisiana researchers say the cometary patterns are best explained by the existence of "a perturber, acting in concert with the galactic tide." Matese said the proposed object should make one orbit around the sun every 4 million to 5 million years. Murray said the object he had in mind would make one orbit every 6 million years, circling the sun in a direction counter to that followed by the nine traditional planets.

The two researchers said they were familiar with each other's work but hadn't taken a close look at each others studies. They acknowledged that their estimates for the mass and orbit of a mysterious object were similar, but couldn't say whether they were talking about the same object. MORE QUESTIONS

How could such a massive object exist so far from the sun? The researchers say a planet or dark star could have coalesced during the formation of the solar system billions of years ago, but more probably would be a passing celestial body that was captured by the sun's subtle gravitational pull.

Another question: Why hasn't such an object been seen? Murray says that even a Jupiter-scale planet could not be observed at the immense distances involved. Matese and his colleagues say that their hypothetical brown dwarf wouldn't have been detected even by the Infrared Astronomical Satellite, which surveyed the heavens in 1983 — but that the yet-to-be-launched Space Infrared Telescope Facility just might be able to pick it up. All this may sound like science fiction, but an expert in the field notes that the hypothesis has been a subject of serious speculation for years. "We've all wondered whether there was something out there," said Brian Marsden, who heads the International Astronomical Union's Central Bureau for Astronomical Telegrams as well as the Minor Planet Center at the Smithsonian Astrophysical Observatory. However, Marsden also expressed some skepticism about the evidence behind the latest research. "I'm not convinced it is not due to chance," he told MSNBC in an e-mail message. "In any case, the data may not be as good as one would like." SPECULATIVE SCENARIO

If the research holds up, it could open the door for renewed speculation on even spookier questions: Some theorists have proposed that the gravitational effect of a massive unseen object in a distant orbit — nicknamed "Nemesis" or the "Death Star" — could set off periodic cometary storms, which would increase the chances of a catastrophic impact with Earth. Indeed, physicist Daniel Whitmire, a colleague of Matese's, who is a co-author of the new research, laid out just such a scenario in 1985 to explain mass extinctions on Earth, such as the demise of the dinosaurs.   Matese also speculated back then about such an effect, but he emphasized that the newly detected object didn't fit the doomsday profile. "This object is not a Nemesis," he told MSNBC. "It does not create comet storms."

He said his proposed object appeared to have an influence on about 25 percent of the long-period comets coming in from the Oort Cloud. Matese noted that theories proposing a correlation between extinctions on Earth and celestial orbits had fallen out of scientific favor in recent years. But he said there could be a "much more gentle" effect that links periodic changes in cratering to the solar system's oscillating motion through the galactic plane. As the solar system moves around the Milky Way's center, it bobs slowly up and down through the galactic disk, Matese explained. The gravitational effects could cause changes in the number of comets sent into the inner solar system, he said.

"We don't know the precise period of that motion" through the plane of the galaxy, he said. "If we discover that it's closer to a 35-million-year period, then a case can be made that it causes periodic changes in cratering."   (end)

The next article is again talking about the same study from the Open University in the UK. It is set in the most unbelievable of publications, a supermarket tabloid. It's what I affectionately call a "Mad Rag." Writers' imaginations have free reign in these things. As Sean Connery's character in the movie "Finding Forester" says,

"The tabloids are what you read for dessert after you finish your serious reading." All supermarket tabloids are all curiously owned by the same company. Since it is coming from this sordid source, they are given free reign to be as alarmist as possible. So from this standpoint, the gravity of the situation is overblown.

Any reference in a supermarket tabloid is unbelievable; so, if your aim is to destroy the believability of evidence, this is where you want it published. In this article they say Planet X is 33 times as big as earth and is going to hit earth and kill us all. As long as you read about it in these sort of publications, they expect you won't ever take it seriously if you come across it elsewhere in a more serious setting. Here facts are freely mixed with exaggerated drivel. (In reality, X is 4-5 times the size of earth and doesn't need to hit earth to cause huge tragedies.)

HUGE MYSTERY OBJECT IN SPACE WORRIES SCIENTISTS!

Some astronomers think it's headed directly toward Earth — & could have catastrophic effects! WASHINGTON, D.C. — Concerned scientists warn that they have discovered a huge mystery object on the outskirts of our solar system, a giant behemoth 33 times the size of Earth and capable of smashing our planet into smithereens!

Although the exact trajectory of the object has not been positively established, some astronomers fear that it is coming directly toward us —- which would mean instant death for every man, woman and child in the world.

"If this object collides with Earth, the impact would be devastating," said an expert. "It would be like a bowling ball smacking into a tiny mothball and blasting into a million particles of dust."

The ominous mystery object, which has been named The Perturber, lies beyond Pluto — which up until now was deemed the farthest of our solar system's nine known planets. Some astronomers say The Perturber might be a newly discovered 10th planet, a recent invader that somehow infiltrated the solar system eons after all the others were born.

Others experts are convinced that The Perturber is a burned-out star known as a brown dwarf —- a once-bright celestial object turned dark and menacing, like an evil twin of our sun.

Whatever the dark intruder is, it is colossal — at least 264,000 miles in diameter, three times the size of the largest known planet, Jupiter. It is so gigantic that its gravitational pull actually distorts the paths of passing comets.

It was by observing the effect on at least 13 comets that John Murray, a planetary scientist at England's Open University, recently discovered The Perturber's existence. U.S. astronomers at Louisana-Lafayette University have confirmed that the frightening object is really out there.

For years, scientists have feared that something like The Perturber could exist. Astrophysicists have long warned that we might someday discover an "anti-sun" or "Nemesis" lurking in the solar system that could threaten all life on Earth.

The unseen object's movements are difficult to calculate, but several observers now agree that it seems to be coming steadily closer. Right now, the object is still billions of miles away, but depending on its rate will arrive "within months," experts say.

"It's as if we're being stalked by something we can't see directly, but we know is out there," said one scientist. "It could strike us next year. "It really is somewhat unsettling to think about it."

Government Posts Picture of Planet X on the Internet July, 2001?

Well, this is the latest addition to this book. Apparently a relatively quiet stir was beginning to brew because of the number of sightings of Planet X from uncontrolled observatories early in 2001.

Now, guess what the powers that be went and did? In order to quiet the rumblings and concerns, a picture of X was posted on the net to save face. Aren't you thankful that they'd be so kind and considerate? Now, do you think they mislabeled X or told the truth straight out as they know it? Think hard now. This is multiple choice and you have two choices. I'm sorry, I'm laughing out loud as I write this. Here's what they called it: "2001 KX76, large reddish chunk of something, Kuiper Belt Object."

I think the disinformation crew stayed up all night, smoking funny cigarettes and worrying about what they were going to call it, then they finally agreed to label it everything they could think of and nothing at all. "Large reddish chunk of something" is my favorite part of the label they came up with for X. This is the part of the bull that is trying to say, "We really truly don't know what it is; believe us, please!"

Because X is still far away, it of course appears small. So, these slick Willie's are trying to say it is, instead, relatively small and close now. Well if this object is close and in Pluto's neighborhood, why haven't we seen it before? Why is it red? Why did it conveniently show up in the exact same part of the heavens as Planet X was spotted a few weeks earlier coming in from the Orion Galaxy? If it is so easy to see now, why wasn't it shown to us earlier? Are you thinking about these questions at all!? Disinformation equals lie. Lie equals cover-up. Cover-up equals conspiracy. Conspiracy is that dirty word we've all been preprogrammed to turn our heads away from and snicker at in disbelief. It's time to embrace and romance one conspiracy in your life if you have any desire to keep yours!

Alas, large approaching objects that seem small at first, appear larger over time as they get closer. A year from now, when X appears larger in size because it's much closer, they are going to have to change their story. Hmmm, what new nonsense will they dish out next? How will they account for their changing story? Their BS is going to smell worse and get funnier as they pile it higher and higher. At a certain point when X is just a few weeks away, the BS will stink so bad you'll just get stone cold silence if you question it.

Ask yourself this question. If this "large reddish chunk of something" were close, small, and backlit from our sun, why doesn't it appear brighter? Why does it look dull red? Here's an explanation you won't hear from NASA. Maybe it's a slow smoldering, brown dwarf star that generates its own dull, red light deep in space from the molten fissures on the surface of its crust? The back cover has their picture of X. Take a good look and read what they have to say — if you like fairy tales. I'll give them credit for their sincere title above the picture. Lies are more believable when labeled with the truth. "LARGE OBJECT DISCOVERED ORBITING SUN."

Large Object Discovered Orbiting Sun

By Larry O'Hanlon, Discovery News

July 3 — The discovery of a large reddish chunk of something orbiting in Pluto's neighborhood has re-ignited the idea that there may be more than nine planets in the solar system. Then again, it could also mean there are only eight planets, depending on your point of view.

What the discoverers are calling 2001 KX76 might be one of the largest "Kuiper Belt Objects" or KBO's, found in the what is essentially a second asteroid belt beyond the orbit of Neptune. Initial reports give 2001 KX76 a diameter of 900 to 1200 kilometers — roughly the size of Pluto's moon, Charon. Pluto itself, it should be noted, is smaller than our own moon.

Because KBOs are believed to have very elongated orbits around the sun they spend a lot of time on dark, centuries-long excursions into deep space. That makes them very hard to find, said astronomer Robert Millis, director of the Lowell Observatory, which was involved in the discovery. Astronomers at the Lowell Observatory have teamed up with colleagues from MIT and the Large Binocular Telescope Observatory to hunt for KBOs on the less remote parts of their orbits. "There are certainly lots of (KBOs) in distant parts of their orbits now, and we can't detect them," said Millis.

The possibility that 2001 KX76 has big brothers and sisters again raises the thorny question of what can be called a planet and what cannot. So far there is no good definition of exactly what is a planet, said astronomer Brian Marsden of Harvard University's Minor Planet Center. The matter only gets more confusing when you add KBOs and free-floating planets discovered outside our solar system.

Historically, Pluto was designated a planet when it was discovered in the 1930s because it was thought to be much larger than it is, Marsden said. The 900-kilometer-wide asteroid Ceres was also considered a planet when it was discovered, until its siblings were turned up and revealed the existence of the asteroid belt between Mars and Jupiter.

If 2001 KX76 is any indication of larger KBOs out there, it might also lead to the demotion of Pluto from puniest planet to king of KBOs, said Marsden. Millis prefers a third alternative: "There may exist a new class of planets."

http://dsc.discovery.com/news/briefs/20010702/kuiper.html#

August 2001

I couldn't even get this new updated version of this book out before the fantasy fellas, or rather the disinformation crew, went and changed their stories about X already. It looks like Planet X has already grown in estimated size from 1200 kilometers or roughly 746 miles across to now as much as 870 miles in diameter. When will this diameter range increase next? They've also added to the long list of funny labels. They're now calling it a "Giant Asteroid," an "icy rock," a "spherical space rock," a "minor planet," and a "Giant space rock," They even say it's "THE LARGEST ASTEROID EVER FOUND." Wow! Good for them. They found that piece of paper with the additional BS labels for X, which they dropped on the floor that night with the funny cigarette roaches.

http://www.eso.org/outreach/press-rel/pr-2001/phot-27-01.html#note3

Pay close attention to these articles. Sometimes the most difficult thing to explain away is just completely left out when piling lie on top of lie. The hardest thing for them to explain away is its color. It's not mentioned in these new articles at all. It is RED! Where do they point that out here? They don't. "The large reddish chunk of something" suddenly has no color.

Asteroids are usually all different kinds of shapes. It also happens to be spherical or round. I wonder why? This object just grew in size in less than a month. They say it is the **"Largest Asteroid ever found now." Well at least they are giving it the up-tick in status it deserves!**

Wasn't this object in Pluto's neighborhood a few weeks ago? Now it is "far out in the Solar System." Are these guys confused or just nervous as hell and trying to say everything they can about it except what they know is the truth?

If X was found in May, why did they wait till August to disclose it? Why did they lay the blame of the initial find on two separate Observatories in different parts of the world, both back in the month of May? Did it take them that long to decide what sort of bs was going to be printed alongside of the picture? Hmmm.

Again, why is there no mention of this object being Red? Is all this sounding fishy to you yet? The astronomer David C. Jewett makes an interesting statement that sums it all up. "Size doesn't matter." So, let's go ahead and ignore the elephant that just walked over the hill in our vicinity. In quotes is one thing Dave had to say, "It fits in with a pattern"; yes, of large red elephants that walk in our backyard.

In the movie, Swordfish, John Travolta's character teaches the audience how **redirecting attention is a powerful tool.** Introducing the new virtual telescope is a simple technique to redirect your attention. Naming Planet X the "Largest Asteroid" is just too alarming by itself. So they turn your mind's attention away from the Giant Red Elephant in our big backyard toward the new pair of binoculars they found it with.

This simple redirecting of your attention away from X to the scope is supposed to help hide several items that have changed from the first disclosure in early July, 2001: First, X's size estimates have increased to the point where it is now being given the label "Largest Asteroid". The latest picture posted of it shows its color to be blue instead of red. In the July disclosure, the statement "large reddish chunk of something" was used. Now there is no mention of its red color and it's suddenly blue. Brown dwarfs are red. If they determined it was time to say it's the "Largest Asteroid," it's going to get a lot more attention than the first disclosure. They determined this was the time to get rid of the red color to stop speculation around its being a brown dwarf with all the new eyes on its size.

In the BBC disclosure/article, there is a heading that is entitled "NO DANGER." Why would they mention danger if it's just an asteroid out in deep space? I tell you, I am completely reassured for my safety now that I've read "NO DANGER." I was worried for a minute until I read this quote: "This asteroid is one of the ones we should be least afraid of," Yeah right. Whenever I see the government(s) describing anything, I develop my concern directly in proportion to the basketful of frivolous descriptions and statements purporting no concern!

Under the heading "NO DANGER," they say X is an "icy rock." Ask yourself this question. If this is an icy rock, why is it red!? Could be a red snow cone maybe? Bloody red ice is what these guys have been drinking trying to cover this up. Myself and one friend could have done a better job of creating a cover story to hide this planet. Jesus, what a bunch of bungling fools! Ok, I'm being judgmental to the negative. Perhaps they were told to print one thing and wrote another, so that thinking rational people could see what they were doing and prepare for Planet X's passage.

The last line in the BBC article says that a new mythological name for the object is forthcoming. May I suggest "Nibiru?" — This is Planet X's name given from ancient Sumerian text. Nibiru translates into "planet of passing." The myth is real.

Giant Asteroid Found Far Out In Solar System

By Frank D. Roylance
Baltimore Sun Staff
8-26-1

European scientists say object near Pluto's orbit is biggest of its kind.

For 200 years the giant asteroid Ceres has held the title as the largest known "minor planet" in the solar system. Ceres is a spherical space rock orbiting in the asteroid belt between Mars and Jupiter. It is nearly 600 miles in diameter, roughly the distance from Baltimore to Chicago. Now a team of European astronomers is claiming that Ceres has been eclipsed in size by a newly discovered object, found near the orbit of Pluto.

The new asteroid could be as big as 870 miles across, according to calculations by a team led by Gerhard Hahn of the German Aerospace Center in Berlin. The team's news release called the data "decisive ... relegating [Ceres] to second place after holding the asteroid size record for two hundred years."

Not so fast, said Brian Marsden, director of the International Astronomical Union's Minor Planet Center. He said an asteroid's size can't be precisely determined without first knowing both its distance and its brightness, or reflectivity - also called its "albedo." The Europeans have securely fixed the object's orbit and distance, he said.

Too soon to tell
But, "it's a little premature for them to boldly come along and give a size, when they're still assuming an albedo." More precise observations are needed, he said.

The new asteroid was discovered in May by a team led by Robert L. Millis, director of the Lowell Observatory in Flagstaff, Ariz. It was temporarily dubbed 2001 KX76. It was found in an orbit beyond Neptune, about 4 billion miles from the sun, in the inner regions of a vast, icy realm of the outer solar system called the Kuiper Belt. The discovery team made a preliminary estimate of KX76's diameter of between 595 and 788 miles, or about half the size of the planet Pluto.

45

More precise calculation

Thursday, however, the European Space Agency Information Center announced a more precise orbital calculation for KX76 using a "virtual telescope" to digitally search for the asteroid on star photos taken years ago.

Coupled to the same assumptions about KX76's brightness, the new orbital data boosted the asteroid's presumed diameter to between 744 and 868 miles. Both KX76's discoverers and the Europeans assumed that the asteroid's albedo lay somewhere between those of another Kuiper Belt asteroid, called 20,000 Varuna, and a typical comet nucleus from that region of the solar system.

David C. Jewitt, an astronomer at the University of Hawaii and co-discoverer last year of the Ceres-sized Varuna, said it makes no difference to science whether KX76 is bigger than Ceres. "It's just a record-keeping thing," he said. The real importance of KX76's size, he said, is that it "fits in with a pattern."

More than 400 Kuiper Belt objects of various sizes have been found since 1992. "And that size distribution probably extends all the way up to Pluto [1,426 miles in diameter] and probably includes Pluto as one of those bodies," he said. And, he said, "it's quite possible there are a few objects bigger than Pluto waiting to be found."

Now here is how the BBC in Europe carried it.

By BBC News Online's Helen Briggs

A giant space rock has entered the record books as the largest known asteroid. European astronomers confirmed on Friday that a distant object seen circling our star near Pluto had broken a 200-year-old record. The previous incumbent was the asteroid Ceres which was discovered in 1801.

People who believe that Pluto is just a minor planet will have more proof now. Lars Lindberg Christensen of the Hubble European Space Agency.

The new object is much bigger, about half the size of Pluto, and is very distant from the Earth. The asteroid was first spotted in May by astronomers at the Cerro Tololo Observatory, Chile. Follow-up studies put its size at 1200 kilometres or more across.

Major and minor

The asteroid is of great interest to astronomers, not just because of its massive size, but because it could shed light on the debate over Pluto's classification as a major planet. Some astronomers believe that Pluto, the smallest planet in the Solar System, is not big enough to be considered a true planet and should instead be called a minor planet.

Lars Lindberg Christensen of the Hubble European Space Agency Information Centre Garching, Germany, said the discovery added weight to this argument. "People who believe that Pluto is just a minor planet will have more proof now," he told BBC News Online.

"No danger"

The icy rock is very distant from the Earth. Mr. Christensen said there was no "apparent danger" that it could ever collide with our planet. "This asteroid is one of the ones we should be least afraid of," he said.

Kuiper Belt Objects

Icy planetary bodies orbit beyond Neptune in the distant region of the Solar System. More than 400 such objects are currently known. They are believed to be remnants of the formation of the Solar System and among the most primitive objects available for study.

The observations were carried out at the European Southern Observatory with the world's first operational "virtual telescope," Astrovirtel. "The concept of a virtual telescope is a highly sophisticated science tool that mines all of the databases to find answers to questions," said Mr. Christensen. The technology allows astronomers to combine data from conventional telescopes with a powerful search tool.

The asteroid has been designated 2001 KX76 for the time being, but will eventually get a real name. As a Kuiper Belt Object, it must be given a mythological name associated with creation.

Estimated sizes, size quotes, locations, or a sampling of bs:

—THE LARGEST ASTEROID EVER FOUND
—900 to 1200 kilometers
—595 and 788 miles across
—600 miles in diameter, roughly the distance from Baltimore to Chicago
—Roughly the size of Pluto's moon Charon
—half the size of Pluto
—distant object seen circling our star near Pluto
—about 4 billion miles from the sun
—one of the largest "Kuiper Belt Objects" or KBO's
—smaller than our own moon
—near the orbit of Pluto
—in a second asteroid belt beyond the orbit of Neptune
—massive size
—size can't be precisely determined
—very distant from the Earth

—beyond Neptune in the distant region of the Solar System

If ever there was a smoking gun of a cover-up, here it is. There happen to be dozens of articles about this 2001 KX76 now. Here is a clip from one. It has now been confirmed that they have records of this object going back at least 18 years. If it was fainter and smaller back then, **now it's clearly larger and getting closer!**

Why do they say it was first discovered back in May 2001 by two separate observatories in different parts of the globe? If you ask them, I'm sure you'll get an ear full of credentialed official sounding nonsense.

Just to drive my point in a little deeper and twist, wasn't it exactly 18 years ago on the last day of 1983 that the chief scientist of the IRAS satellite announced they'd found the 10th planet to 6 daily newspapers? I think the Keystone Cops are in charge of this conspiracy. They admittedly have 18 years worth of data on 2001 KX76, and the 10th planet was officially found 18 years ago.

The first two articles back in early July, 2001 mention its being red. There's a picture that shows it to be red to go with the article. This red object stands out in a sea of blue ones like a sore thumb.

The half dozen or more new articles in August don't mention or show it being red at all. Now there's a new picture to go with the new disinformation. When my son was growing up, he described everything to me as happening suddenly. Well **suddenly**, 2001 KX76, or whatever nonsense name they give **it, is BLUE!** Too many slow-smoldering brown dwarfs are red perhaps? Maybe they had to change the color in midstream to deflect the brown dwarf questions?

In the entire list of speculative names they gave this object, not one time could they say it was possibly a brown dwarf star. They don't sound too sure of themselves, so why not mention brown dwarf as a possibility also? Is it a bad word? Now all of a sudden we have the "largest asteroid ever recorded," and it's growing, changing colors and shifting locations.

X is red. It's just too difficult to explain away the color and not mention the possibility of its being a smoldering brown dwarf. So why not go and doctor the picture to make it look blue? Sounds good to me. If somebody asks about it, we've got a whole new paragraph of double-speak ready and waiting to explain away the new color. At least now we won't have to mention the new color in the latest articles because most other objects alongside of it are the same new color blue. It's difficult to bite my tongue and not say what I think of these people behind this cover-up. Here's the piece that admits they have pictures from 18 years ago.

> *The latest is this, the label 2001 KX76 has been used to identify at least 2 seperate objects near where Planet X is coming in from. The orbital elements have changed. On the back cover you can see the original double red image from July 2001, one is Planet X. So, this is one of the techniques that*

NASA is using to hide X. To find the latest coordinates, used successfully several times this year, 2001, to site X, go here: http://www.zetatalk.com/usenet/use90207.htm

The search was successful: the astronomers were able to find several photographic plates on which faint images of 2001 KX76 could be identified - some of these plates had been obtained as early as 1982. The exact sky positions were measured and with accurate positional data now available over a time span of no less than 18 years.

Intuitives, Prophets, Legend, and Folklore

When I first started studying intuitives in relation to earth changes, I had no idea that there was an exact date (plus or minus a couple weeks) for the worst of these events. Neither was I aware that there was a definitive cause for them since so many different theories abounded. I couldn't have imagined that parts of the government and its select controllers could have known all about this for decades either. There are many different prophecies that have earth change themes to them. The cultures and times that these legends, folklore, prophets and intuitives speak from vary as widely as the individuals' explanations for how they came upon the knowledge. All curiously have gotten the same basic message and point to around the same time period for these events to happen. Some of the ancient texts that this generated from may not have come from true prophets at all. If they had good information that stated the regularity of the orbit of Planet X and knew the approximate date of the previous passage, they could have figured out when it was going to return and created prophecies that would end up being accurate.

There are, however, some true prophets, futurists or intuitives that are able to see future events with a high degree of accuracy. This information originates from a collective consciousness that the US military government has openly admitted using, when they've obtained information through their "Remote Viewing" experiments.

Some intuitives are just too accurate to be denied. I believe that they should be looked at collectively. Of the many prophets or intuitives that have mentioned earth changes, I'm going to give you just a sampling of a few of them. Some might be familiar to you and others not. On the other side of the world, I'm sure there are

49

several that I'll never learn about. Each culture and time have their share of them. This in itself gives rise to a greater intelligence that has a quiet unseen hand in all affairs.

Science is a battleground for who's right and who's wrong, according to the latest popular thinking. What speaks to the general public *en masse* more deeply than the latest experts' ever-changing views? It's the Prophets, folklore and legend, that have always touched the hearts and minds of the masses more deeply than continuing evolution and the endless stream of scientific theories. I myself started my study of this earth changes subject matter through the eyes of the prophets. Then my knowing and certainty converged, when I found all the scientific evidence and ancient history from previous events clearly stating the reason and regularity for the changes. All the pieces of the puzzle started to fit into place as to why this evidence was being suppressed and by whom.

Prophecy has the emotional impact that opens people's eyes *en masse* to make a change to safety unlike science or conspiracy. So, before I present more data, I will go back to my roots on this subject. I'm now going to concentrate on the high impact, heart-wrenching gift of the gods. I push forward here with messages that came through intuitives from wherever your personal philosophy can imagine they're from. These segments may allow one to emotionally realize, now is the time to take action to save your life and the ones you care for so dearly. Follow your heart and move to a safe place, and don't listen to the ones who can't see how close to the changes we are. It's your life, not theirs.

The collective mind, from which the intuitive's information originates, is not confined to space or time. It can therefore see what lies ahead of us on the linear time continuum; our perceptions focus on but a short piece of it, which we call "the present." Now if you become a dedicated student of intuitives, you will find that there are 1000's of earth change prophecies from an untold number of individuals, groups and cultures around the globe. This sheer volume of messages would not exist if Earth didn't experience regular cataclysmic changes. There would be no need for our collective mind to be sending these messages to us through our most perceptive individuals throughout the entire world if this were not the case.

The most élite and the sharpest on this planet are mainly not the ones this society looks up to. It's the spiritually connected individuals that are truly the very best and brightest our society has to offer.

When The Comet Runs, by Tom Kay

I am a follower. The truly unique person is the one who blazes the trail. If you truly sense that the impact of prophecy is more powerful than all else on the human psyche, then, before you check out anything else read carefully "When The Comet

Runs" by Tom Kay, published by Hampton Roads Publishing Company Inc. He tied together prophecies, Planet X and Earthchanges long before I knew the connection between them.

Mayan Calendar vs. Gregorian

Before I give a sampling of visions from a few intuitives regarding earth changes, let me first set the stage here, by relaying how the Mayan calendar possibly meshes in with our Gregorian one.

The Montel Williams show had a noted California psychic **Sylvia Brown** as their guest. It was taped in January, 1997. In the midst of giving predictions for 1997, she mentioned that 2000 AD actually took place ten years ago (1991?).
Therefore:

Gregorian	Mayan
1991	2000
1992	2001
1993	2002
1994	2003
1995	2004
1996	2005
1997	2006
1998	2007
1999	2008
2000	2009
2001	2010
2002	2011
2003	2012 AD

To summarize, 2003 AD (Gregorian calendar) translates into the Mayan calendar's 2012 AD. Remember that the Mayan calendar ends Dec 21, 2012, completing a cycle or signaling a catastrophic end of the world as we know it.

Intuitive Message from Spirit # 1

Ruth Montgomery, *Strangers Among Us* (1979)

Her guides told her to look for an absolutely unavoidable shift in the earth's axis and crust around the beginning of the next century. Regardless of any thought or deed by those in physical form, and despite all the efforts of scientists and engineers to avert it, the shift will occur.

The shift will have its warnings. The revolutions of the earth will be slowed in the last years of this century. The weather will be worsening in most areas. The storms will become increasingly violent, with heavy snowfalls, strong gales, and increased humidity. There will be rumblings beneath the earth, and the trees will sway. Shortly before the actual shift, there will be two specific types of warnings. The eruptions of ancient volcanoes in Mediterranean islands, South America and California will result in pestilence, and shortly thereafter earth tremors of major proportions, affecting wide land-masses in Northern Europe, Asia, and South America, will provoke tidal waves of monumental scope. These, then, will be the forerunners of the shift itself, and for days and nights beforehand the earth will seem to rock gently, as if soothing an infant in its trundle bed. They said some will recognize this as the time to remove themselves quickly from the seacoasts and other exposed places, and, while that exodus is occurring, there will be increasing earthquakes and volcanic eruptions in flat areas that had previously shown no sign of cones. Some will remain despite the alarms, disbelieving that a shift will occur; some, deciding that it is a good time to return to spirit, will refuse to leave their homes.

The earth will hesitate in its orbit prior to the shift. In daylight areas, the sun will seem to stand still overhead, and then to race backward for the brief period while the earth settles into its new position relative to the sun. In nighttime areas, the stars will seem to swing giddily in the heavens, and as dawn breaks the sun will seemingly rise from the wrong place on the horizon. Those who are capable of reaching safety will see the earth's surface tremble, shudder, and in some places become a sea of boiling water. Water will pour over and cover many areas of our present continents. Simultaneous explosions beneath the earth's crust will bring new land above the surface of the waters as other areas are swallowed by the sea. We ask that you picture a giant wave, higher than a ten-story building, racing toward shore. Impossible to escape it, so in that moment of terror it is well to put aside fear and think only of the good that is to come by passing into spirit. Conquer fear, and one has risen above the battle. For the survivors there will be anguish and heartache, but also exhilaration in having withstood the ferocity of nature, and most of those remaining will feel that god saved them for a purpose.

There will simply occur drastic climate alterations. Some cold places will become warm and some warm places will freeze over. Countless numbers of humans, animals, and birds will freeze to death where the new poles settle. Ferocious and devastatingly high winds will sweep away most above ground structures. New York City will vanish. Florida will scarcely survive, except as scattered islands. The southern states bordering the Atlantic and Gulf of Mexico will be drastically altered, including parts of Texas. In the west, the remainder of California will disappear beneath the broiling waves. Canada will be in a warmer latitude and much of it will be relatively safe from sinking or destruction by tidal waves, not a bad place to be if survival is your choice. Washington D.C. will be devastated, but not totally destroyed being near the mountains. Government workers will carry on in the previously prepared shelters there beneath and within the solid rock. Virginia Beach will strangely survive, as most other seaside resorts disappear. Whole areas of the eastern and western United States will be deluged. Hawaii slides into the sea. Survivors will be thinking of moving to inland areas before the shift. We do not mean that the elderly should flee the sunbelt. Rather, they should give serious thought to whether it is worth the effort to make drastic alterations in life-style in their declining years. Civilization will diminish as the earth changes wipe out or render useless hydroelectric plants, housing developments, office buildings, skyscrapers, dams, refineries, telephone lines, dock facilities and more. Cities will be in utter chaos. Suburbs will be the first areas thronged by weary refugees from metropolitan centers scavenging and stealing to keep their bodies alive.

Her guides said that at the time of the writing, preparations were already well underway for the problems of assimilating a huge influx of souls crossing over simultaneously. Ultimately, whether you find yourself with or without a body, she wants to point out that our souls are immortal, and we should not worry about surviving as such. She states that standing up to the stresses and helping others will forward your soul's progress. She says that the shift will also eradicate some of the evils of the present age along with sweeping clean the beastliness and cupidity that surrounds us. Serenity and love toward your fellow man are important to the new world she sees ahead. She, like many authors, sees a brighter future for those that are left and sees a flowering of a civilization in which the best of man's instincts are given full range. The shift will eradicate mind control and closed societies because all peoples will be working together for survival.

Those who fear the event should think of the exciting movies they have watched, such as Star Wars, and realize that they will be privileged to witness one of the greatest events of all ages. Those who survive will "dine out on the story" as the phrase goes, for the remainder of their lives, telling the new arrivals and grandchildren about it and pointing out where there used to be land and now is water. Think of the thrill for those hundreds of miles inland, to peer out the next morning and discover that they are now living beside the sea! There will be much to commend this experience, so do not think of it as a diabolical era. Her guides say millions of people will survive the shift in their physical bodies. It will be an interesting and fascinating time that's full of challenges and opportunities.

There are also records from many great civilizations that have come and gone before ours which will be found after the shift according to Ruth.

Her guides have said this shift is a cosmic event, something big to study, watch and help out with in our part of the universe. Thus UFO and alien sightings will be on the increase. Apparently there are also many terrestrial organizations that have known about and have been preparing for this event for decades. These people have planted fruit and nut trees and have stored grains for future needs. Some are already teaching basic survival skills, and some have stored their documents in safe places. There will be safe areas. As the time approaches, these will be made known through Brotherhoods and inner awareness, so that those wishing to survive in physical being will be aided in going there. **<u>Acquiring skills that we can share or barter with others is an especially valuable preparation.</u>**

For many years now a quiet trend has emerged. Some are learning proper use of the soil, others are acquiring medical training or midwifery skills, learning how to work with fabrics, leather, and skins, developing herb gardens, drying fruits and vegetables for preservation, fashioning pottery from local clay, and discovering how to harness water power to run machines off the grid system. Everyone who wants to survive would do well to learn a few basic skills that can be shared. They would also be wise to gather such simple equipment as needles, sharpening stones, knives, axes, fish hooks, hammers, and other hand tools that are not electrically operated.

All of us are aware of the catastrophic conditions that often result from a massive power failure in our cities, with the resultant looting and rampaging through the streets, the food spoilage, and the lack of fuel. We know what havoc can occur when an unusually heavy snowstorm paralyzes traffic, halts deliveries, and blocks garbage collection. If pandemonium is created by such relatively isolated occurrences as these, it is mind-boggling to think of the worldwide chaos that could result from a shift of the axis. Most distribution and transportation systems will collapse, and if people are without basic skills, and without knowledge of how to work together, they will not survive.

The travails and heartaches predicted for the next few decades will not have been endured in vain if the Guides correctly foresee the golden age to follow. That period, they say, will see the commencement of the millennium, as foretold in the Book of Revelations. The twenty-first century will be a time of beatitude. All beings will work together to rebuild the shattered earth and renew its fertility, which will have been drastically affected by the alteration of the earth's poles. New lands emerging from the seas will at first be saline, but will soon begin producing quantities of exotic foods, and a new type of agriculture will emerge. Some areas that once were fertile will have become arid deserts or sea bottoms, and others which had been overproducing due to artificial fertilizers will wisely be left fallow for a time. The ability to read minds through mental telepathy will spread rapidly, outmoding some communication systems, and a peaceful age will flower. In the twenty-first century arms will become virtually unknown. After the shift, most will be cleansed of evil-thinking in the common desire to survive and rebuild. The shift will so shock humanity that man will devote himself to helping others as himself, and good thoughts will project to such an extent that it will seem like a new race of human beings. The period of chaos and cleanup will eliminate many who were sickly and had relied on pill-popping instead of their inner reserves, as well as most of the freeloaders, thieves, and the greedy. The majority will be healthy, strong, loving, kind, and sharing, the traits that can withstand calamity and rebuild. The vast destruction and turmoil will have been necessary in order for the next step in man's development to begin. The finding of one's inner self to survive, after losing everything material, will forward man's spiritual progress.

Message from Spirit # 2

Deborah Harmes, Ph.D. has had with her since childhood a being that is known as the Dreamkeeper. The following is from her book "The Dreamkeeper." It is a group being that describes itself as the keeper of dreams that bind all existence together. The principal messages are responsibility, free choice, and honesty. Plus it stresses one ability as paramount, to find your individual truth for all matters by looking within and not to any other person or organization unless it rings true to YOUR heart. The Dreamkeeper is a non physical being that never was nor will be physical. It shows Deborah visions and communicates through automatic writing and channeling.

The long prophesied changes for your world are here now and there is no escaping that fact. You will begin to read tiny and uncomplicated reports in your newspapers that will introduce you to the idea of magnetic shifts, strange patterns in magnetic energy, or magnetic storms in space that affect the technology of Earth. This is a means for preparing the population of the planet for what the scientists know is already an irreversible activity. The weather patterns you have observed in the last few years are not the weather patterns of a mere decade ago. Can you ever remember a time when there were so many floods, droughts, tornadoes, hurricanes, earthquakes, and volcanoes all happening in the same year?

Those of you who report a feeling of rocking back and forth, as if on the deck of a ship, are simply noticing the instability of the magnetic fields that surround your polar alignment. When all is dark and all such creatures should be resting prior to their dawn chorus, they too sense the magnetism "offness" of the atmosphere and cannot rest. Is it any wonder then that if the animals are already sounding the alarm, you too are feeling a sensation of discomfort?

If you picture the core of the Earth as the large iron sphere that it is, the magnetic pull of celestial objects is increasing the instability of the Earth's rotation and is also causing a flux in the magnetic atmosphere of the planet with each passing year. Celestial objects not only affect your gravitational and magnetic fields, but they influence your weather patterns. You will find more and more evidence of severe weather that suddenly appears without warning and causes great loss of life and property. These wild outbursts of destructive weather events will increase with each new year. Your weather predictors are only now beginning to understand that the patterns of weather that they had always been able to predict in the past are no longer valid. This is due in part to the influence of these celestial objects passing in the vicinity of your "Earth aura."

The inner and outer self of the planet is increasing in temperature and that is the reason for the melt-off of the polar icecaps and the rising level of the oceans throughout the world. We cannot say that these things will be reversed. They are already well under way and are now irreversible. This is an easily verifiable fact, and the change is measurable. There are no beaches anywhere on the planet that have not been affected by this rising water — and it will escalate with each year.

Part of this can be looked on as a cyclical event. Any geologist can tell you that the position of the North and South poles have moved many times in ages past — and when it happens, it is so sudden that it can bring on events such as the sinking of large land masses or the melting of polar caps within the span of a day. The great mammoths of the place you call Russia were frozen in place with the warm-weather flowers and grains still in their mouths.

The upcoming years will be filed with so many non-stop disasters of one kind or another that you will begin to notice a great many people leaving the planet this way. The Earth is attempting to give you an advance warning of what is coming when she shakes off some of the energy build-up through an increasing number of seismic and weather events. Your scientists and specialists are reluctant to let you know that they have no solutions to offer. But there is soon to be a time when you will look back on the current situation and realize that these were merely minor energy releases. The world will soon be a very different place for you. These big movements and severe weather will escalate until the one fine day when the Earth will toss, roll, shift axis and forever change the appearance of her surface as we know it. These types of events have happened in cycles that have been repeated many times in ages long past.

We, and anyone else who is a responsible "prophet" of the times, are not telling you these things to create fear and panic, but rather so you may be in a state of mental and spiritual preparedness. Many of you do not choose to survive the upcoming changes, and an equal or greater number of you choose to keep hanging onto human life. We encourage you to prepare and do the personal spiritual cleansing that will allow you to be at peace if you are not in a "safe place" during the great shift. The largest of the geological events we see in **2003** or shortly there after.

What will you and those around you do to survive when the electrical, gas and oil power that you are so dependent on is unavailable? What will you do without clean and drinkable water? And how long will you be able to cope without a means to cleanse your bodies? If you have dealt with the fact that the Earth Changes are inevitable, what of the time leading up to that? Have you examined how you will manage until the Earth's Dance and the Transformation occurs? As the weather and wind grow progressively more fierce, where will you go to stay sheltered? Do you imagine that your government relief agencies will be able to provide enough food and housing for large groups of displaced people in more than one location at a time without bankrupting themselves or being forced to turn people away?

Another means of exiting the planet in the future is by eating and drinking contaminated food or liquid. It is essential that you have a source of stored food that is safe to ingest. It is time for you to make preparations to store several months of food in case of long term loss of crops. And if you are able, perhaps this is also the time to learn to grow the essentials of what you could to survive on your own. To ensure that you can take care of one another, make preparations and stop just thinking about doing it.

Prior to the shift, starting now, another effect of the weather is the lack of certain foods or the increased cost of the remaining crops. As the severe weather of all types sweeps across your planet, many crops that are essential parts of your diet will be destroyed. And in some cases, farmers may just recover from one weather disaster and be replanting or ready to harvest a subsequent crop when another weather system destroys the follow-up crop. How are you going to deal with this when the food that you had always anticipated having available is no longer there?

We advocate the creation of smaller, intentionally created communities that are based on shared spiritual and social belief systems. These types of communities are frequently in more rural locations. Most of the existing and forming communities are well aware of the weather and geological challenges that they face and have accordingly stored food, water, medical supplies, and building supplies. It is a way of thinking that you would do well to emulate.

Canada will be known as one of the safest and least impacted locations on the planet. The weather will become more temperate than what is known there currently.

The United States will be strongly impacted by the Earth Changes as most of both coastlines, east and west, succumb to ever-rising oceans. We see that part of the west coast of this country will actually begin to sink into the sea long before the actual pole shift. This has already begun although the occupants of this area are thus far unaware of this fact. The great volcanoes in the states of Washington and Oregon, along with as-yet-known ones in California, will roar into wakefulness to signal the movement of the land masses into a forever-changed state of appearance. There will be no advance warning. As the poles begin to realign, there will be no place within 100 miles of either coast that will not be washed over with the rising ocean.

The pole movement that we have described, combined with an activation of a fault line that parallels the Mississippi River, will cause a dumping of the water masses that border the country of Canada and a new and rather large inland sea will be created all along the line of the former Mississippi River. This river will simply cease to be as it is absorbed into the inland sea and the many states that had bordered this river will either be reduced in size or eliminated altogether.

Most of Louisiana, Mississippi, Alabama and Florida will simply cease to exist, and over half of Georgia and Texas will go into the ocean as well. We also see a large chasm opening in the mid-upper part of Texas as the movement of water under this land mass causes it to cave in and suck in all the land and towns around it in the process. (There is much more information concerning different areas. I could write a whole book covering what all the intuitives say about each part of the world. That is not the focus of this book. E-mail The Dreamkeeper if you wish more from this source at dharmes@bellsouth.net. I'll be mentioning others later.)

It will be many years before the dust and debris completely clear from your skies following the pole shift and resultant eruption of all volcanoes world wide in a simultaneous clearing of energy. For that reason, you must be prepared to live without freshly grown food for that time period or be limited to only what root vegetables will grow in extremely low light conditions.

The great prophets throughout time have correctly reported this time period as that of the end of life on this planet as you know it. Yet this does not mean the end of all human life. On the contrary — it is the beginning of a new and better form of life. This is your destiny, your future, your return to the reality of your being.

Message # 3

In the late 1970s, a group of students and supporters in Great Britain asked intuitive **David Jevons** to request information from **his Ramala Master** in the spiritual beyond concerning the Earth Changes that had been predicted by so many seers to take place near the end of this century. A pronouncement with respect to the "Fiery Messenger" was voluntecred in the reading. The following is from *"The Wisdom of Ramala,* chapter entitled, "The End of The World?"

I know there are people, some spiritually motivated, who believe that this event will not come about and that it will be prevented either by the intervention of some great master or by Humanity reversing the path upon which it is now set. I would ask you to remember the impact of the last great impulse of the Christ energy, of the Master Jesus who came on the earth two thousand years ago. Consider how long it took for that energy to become an effective force on the plane of Earth, even after that Master's great sacrifice. Even if the Christ energy were to return at this time, it could not move Humanity from the course on which it is set. That is why the prophets and seers of old could make their prophecies with such certainty. To ground cosmic knowledge on the earth, to manifest it through the cycle of human evolution, takes time. The spiritual consciousness needed to save this World cannot be grounded in the time that is now left before the Earth Changes. . .

I believe that the major Earth Changes will be initiated by what I will call the "Fiery Messenger." There is even now a star of great power proceeding towards our Solar Body. The star, at this moment, is invisible to the human, or even telescopic eye, but it is set on a path which will bring it into conjunction with our Planetary System. As it passes by, it will affect the motions of all the planets of our System, therefore will bring about changes on the surface of the planets themselves. The effect of the passage will be to set in motion the Earth Changes that are prophesied. Various lands will sink, others will rise. . .

So I say to you now, as I said to you five years ago, that these Earth Changes are coming. They cannot be avoided. They are part of the destiny of the earth.

Is it not strange how Humanity finds it difficult to plan beyond the year 2000. It is almost as if the end of the century is the ending of a cycle. Now I'm not saying that this is the year when these changes will come to pass, but certainly the final ending of the Piscean cycle will take place around that time. This therefore gives you two decades in which to prepare yourselves (written in 1979), to prepare your lifeboats, to establish your true values, to shine your light and to prepare for the ending of your world. I hasten to say your world, not the world, for it is your world that must change, not *the* World... The divinity of planet earth will not be extinguished by any human action.

Although Humanity has the power to destroy itself, it will destroy itself not by nuclear explosions, not by destroying the planet which it abuses out of ignorance and greed, but by destroying its own soul.

Question: The star that was spoken of, has it passed through this Solar System before?

Answer: Yes.

Question: Is it Halley's comet?

Answer: No. It is far bigger than that.

The changes of which we talk now are the Earth Changes that are associated with the introduction of the Aquarian Age. It is essential that those of you who live at this time of transition should understand the nature of and the purpose for these Earth Changes.

Remember that the symbol of the Aquarian Age is the phoenix. The phoenix is the mythical bird which consciously sacrifices itself on the cosmic fire, releasing its old form in order to come forth purified in the new. Does not the phoenix symbolize the desire within your own spiritual being for the purification of the earth to take place so that the old human form can be cast off and the new Aquarian form may come forth?

As you look around the world today, you cannot help but notice the increasing tempo of human conflict all over the globe as both nations and individuals oppose each other for political, ideological, and religious reasons. But humanity is not only suffering from an outer level through famine, earthquake, disease, and war but also on an inner level through its lack of spirituality, its self-centeredness, its greed, its concern only for self at the expense of its fellow human beings and the other Kingdoms of the earth (animal-vegetable-mineral). All these events bear witness to the approach of Armageddon and the ending of the Age. Humanity needs to be purified. Humanity needs to experience the cosmic fire of purification in order to come forth reborn in the Aquarian Age.

If I say that a cataclysm is coming, do not think that disaster looms nigh, that the purpose of your life is limited, that everything is to be destroyed in it and, therefore, that there is no point in pursuing the aims of your life. A cataclysm does bring change, but you are forever changing. In every hour of every day, as you live in your physical bodies of matter, you are changing, and you will continue to change and evolve until the moment of that cataclysm. For there will be death in it but, as you know, death is only another form of change. Therefore death in a cataclysm does not mean the extermination, the ending, of life: it is, rather, a rebirth. I would therefore invite you to regard a cataclysm not as an ending, not as a finality, but truly as a beginning. I would ask you to look at the cataclysm which is to come around the end of this century, not as the ending of an Age but as the birth, the dawning of a New Age.

Cataclysms are your Creator's way of ensuring that the continuing Plan for evolution of this Earth is carried out. They are as natural as the other changes which Man can observe on the surface of this Earth—the birth and death of Man, the birth and death of Nature... Everything in matter is in a continuous state of change. It is up to Man's consciousness to interpret and to recognize the purpose of that change, and then evolution will take place...

Forces beyond your control acting both within and outside the realm of this planet set in motion the mechanism for initiating a cataclysm... When such influences are brought to bear upon the Earth, magnetic forces within the Earth react to these vibrations and thereby trigger off a state of fluidity in the Earth's crust which allows movement of the land masses on the surface.

When these cataclysmic changes take place, large tracts of land are moved around like pieces of a jig-saw puzzle. Large portions of the Earth's surface rise and fall, appear and disappear. This is what confuses your geologists today, for they look at the surface of this Earth as they now see it and try to deduce its whole evolution from just one small portion of its present surface.

I am asking you, therefore, to understand the need for, and the purpose of, a cataclysm, to see why Man has to change, to see why Man must change... for the Age of Aquarius is to be an age of great evolution... In this New Age Man will progress and evolve beyond your wildest dreams. The Earth will become what it should be:

a vibration of Universal Love fulfilling its purpose in the Solar Body. It will be giving out its emanations, not to this Solar Body, but to Creation Beyond.

Many men have prophesied that a cataclysm is coming. Over the last one thousand years, many seers and prophets have spoken of this event. Before you dismiss these prophecies as the warnings of cranks seek to establish why they were warning you. They have long since died: the warnings were not for them. The only motivation they had in making their prophecies was to foretell what was to come. They prophesied so as to warn a race of men which would be far removed from them both in its way of life and in its evolution. You may disregard the voice of God at your peril.

Message # 4

Edgar Cayce

"When there is the first breaking up of some conditions in the South Sea (that's South Pacific, to be sure), and those as apparent in the sinking or rising of that's almost opposite same, or in the Mediterranean, and the Etna area, then we may know it has begun."

Question: "Will there be any physical changes in the earth's surface in North America? If so, what sections will be affected, and how?

"EC: "All over the country we will find many physical changes of a minor or greater degree. The greater change, as we find, in America, will be the North Atlantic Seaboard. Watch New York, Connecticut, and the like."

From Reading #3976-15 (Given on January 19, 1934) As to the changes physical again: The earth will be broken up in the western portion of America. The greater portion of Japan must go into the sea. The upper portion of Europe will be changed as in the twinkling of an eye. Land will appear off the east coast of America. There will be the upheavals in the Arctic and in the Antarctic that will make for the eruption of volcanoes in the Torrid areas, and there will be shifting then of the poles - so that where there has been those of a frigid or the semi-tropical will become the more tropical, and moss and fern will grow.

For many portions of the east will be disturbed, as well as many portions of the west coast, as well as the central portion of the U.S... lands will appear in the Atlantic, as well as in the Pacific. And what is the coastline now of many a land will be the bed of the ocean. Even many of the battlefields of the present will be ocean, will be the seas, the bays, the lands over which the New Order will carry on their trade as one with another.

Portions of the now east coast of New York, or New York City itself, will in the main disappear. The southern portion of Carolina, Georgia-these will disappear.

The waters of the lakes will empty into the Gulf, rather than the waterway over which such discussions have been recently made. It would be well if the waterway [St. Lawrence Seaway] were prepared.

Message # 5

Gordon Michael Scallion

The following is part of a feature article entitled "The Blue Star" in March, 1993 issue of the *Earth Changes Report*:

Prior to the Earth Changes and major shifts in consciousness, spiritual forces send guidance to prepare us for these changes. This time period is known as Tribulation —- a seven-year cycle. (The 7 years cycle that many talk about is the period of time right before Planet X passes. It is close enough at about 7 years away to magnetically effect the heating up of the core of Earth to start the relatively minor changes in weather and quake patterns before the passing and shift.— MH) On the physical/mental level these warnings are perceived within. As changes increase, outward signs occur, such as spiritual manifestations and signs in the heavens. One such heavenly sign is the Blue Star.

I am now shown the Earth shuddering everywhere. There are winds. I am watching large land masses which appear to be the North American Plate and the Pacific Plate. They are moving, perhaps half of the plate structure of the Earth, shift as if in a single moment. They do not go down or up, but rather, they slip. The movement from my perspective doesn't seem far, but it might be something akin to twenty-five or thirty degrees of slippage.

I watch certain land masses that were warm become instantly cold. I see animals, grazing animals that look like herds of cattle, frozen in their paths. I watch other areas that are mile-high with snow melting. I realize that it is the Antartic. As I am watching, there is a time lapse which I would assume would be weeks or months. I watch Greenland, the ice is melting so fast that water levels throughout the world are rising.

The water is moving in so fast that new seaways are made. I can see inland seas in the United States, I can see a river running from the Great Lakes to Phoenix. I can see that the St. Lawrence Seaway has become a large inland sea. The Mississippi divides the United States in two. Europe has become a series of larger islands and most of northern Europe is under water.

At the same time I am seeing other land masses thrust up from the bottom as a result of the shift. I see huge land masses in the Atlantic and Pacific thrust up, even though the melting has raised the water level. I move around to see how many tectonic plates there are. After counting I find their number has increased from twelve to twenty-four...

Now, imagine a star—a blue star, that moves through the heavens at regular cycle, such as a comet. This star has moved through the heavens at various cycles, passing slowly sometimes, pausing other times, and appearing to remain still at other times.

It has visited the earth many times. At the time it came to fulfill the prophecies of old. This same star also visited the earth years ago to warn the world of the coming flood—the sinking of Atlantis.

Gordon Michael Scallion has future maps of how the world will be after the shift. I strongly recommend his if you're interested in finding safe places. He also has a newsletter and a book called "Notes From the Cosmos."

Message # 6

There's a good chance you've heard of this prophet before. Here it is from none other than **Nostradamus**. Michel De Notredame is The Man Who Saw Tomorrow.

After great misery for mankind an even greater one approaches,
when the great cycle of the centuries is renewed.
It will rain blood, milk, famine, war and disease.
In the sky will be seen a fire, dragging a tail of sparks.
There will come a dreadful destruction of people and animals.
Suddenly a vengeance will be revealed, a hundred hands,
thirst and hunger, when the Comet will pass. {II-62}

The dust that follows X is red. As Earth passes through its tail, prior to the shift, a red dust so fine it's unseen turns waters red as it rains down. This explains the raining of <u>blood</u> metaphor. During X's approach, its magnetism causes weather and seismic disturbances which result in droughts and <u>famine</u>. <u>War</u> is a possibility because famine can trigger people in need to try to take their neighbors' goods. I also understand there falls a white sticky substance that is also a part of the passing. This will come forth out of the volcanoes around the globe simultaneously spewing forth hydrocarbons. It is this volcano-based hydrocarbon that resembles a <u>milky</u> web on plants, when this rains down from the sky. Conditions such as in the middle of a major cataclysm are always accompanied by <u>disease</u>. Planet X is lit up from our sun and looks to be a large comet with a tail when it goes by in the sky (<u>fire, dragging a tail of sparks</u>). There are also high altitude lightning storms that cause quite a light show that accounts for the sparks. By the time it can be seen with the naked eye, its strong magnetic influence will have already caused several major events. (<u>After great misery for mankind, an even greater one approaches</u>.)

Message # 7

This is a **combination of prophets** that I will offer as one.

Names for the day Planet X passes from biblical sources would be: "The Last Judgment," "The Great and Terrible Day of The Lord," "Armageddon," "The Three Days of Darkness" (Catholic Version), and the "Apocalypse."

For there will be great distress, unequaled from the beginning of the world until now—-and never to be equaled again . . .Immediately after the distress of those days, the sun will be darkened, and the moon will not give its light; the stars will fall from the sky, and the heavenly bodies will be shaken... (Matthew 24:29)

"And the third angel sounded, and there fell a great star from heaven, burning as if it were a lamp, and it fell upon a third part of the rivers and upon the fountains of waters; and the name of the star is called Wormwood. And a third part of the waters became wormwood; and many men died from the waters because they were made bitter. And the fourth angel sounded, and a third part of the sun was smitten, and a third part of the moon and a third part of the stars, so that a third part of them was darkened; and a third of the day shone not, and the night likewise." (Revelation 8:10-12)

Impending Signs: Increasing violent disturbances on land, at sea and in the air; earthquakes, hurricanes, tornadoes, windstorms, cloudbursts, breaking of dams, overflowing of streams and seas, huge tsunamis, floods, unusual "accidents" taking many lives. Famines, epidemics, destruction, destitution, failure of crops — water and crops will change, becoming more "polluted" and less nourishing. Revolutions, downfall of governments, dissensions, wars, confusion in high places, lack of respect for authority, treachery, corruption, brutalities, atrocities. Immorality, lack of charity, heartlessness, indifference and lack of concern for our neighbor, people turning against each other. A cross will appear in the sky.

(This is Planet X Crossing the horizon with its long twisted tail of debris caught in its strong gravitational wake.—MH)

All will see it, but some will reject it. It will be a sign that the final events are near.

Immediate Signs: The night will be bitterly cold; the wind will howl and roar, then will come lightning, thunderbolts, earthquakes; the stars and heavenly bodies will be disturbed and restless. There will be no light, but TOTAL BLACKNESS. Hurricanes of fires will rain forth from heaven and spread over all the earth. Fear will seize mortals at the sight of these clouds of fire, and great will be their cries of lamentation. Many godless will burn in the open fields like withered grass. 75% of the earth's population will be lost. Source: Blessed Anna Maria Taigi (Rome), Padre Pio (Italy), Pere Lamy (France), Elizabeth Canori-Mora (Rome), Saint Hildegard (Germany), Marie Martel (Normandy), and many more.

Message # 8

Mary's Message To The World
By Annie Kirkwood

The planet is already beginning to have some violent reactions to inner stirring which is occurring deep in its core. The spinning and churning of the core will also have an effect on its rotation. A wobbling effect on earth's spin will cause the ocean currents to flow in different patterns. These changes in the ocean currents will affect every shoreline. The fish and mammals will be confused and some will act different. The tides will become out of sync. Many new records will be set; many unusual events will be reported... Everything will commence to be strange to you on earth. The world is changing, and there is nothing you can do about it. Man cannot control the size and frequency of the storms which will come out of the oceans. Man is helpless against earthquakes, which will be increasing in size, intensity, and frequency. Volcanoes will come out of nowhere. Old volcanoes will become active and alive with fury. These are just the beginning of the coming events. This intensity of the vibrations of the magnetic field will cause many new and powerful reactions on earth. You have just begun to see the fury of these storms. The tidal waves will also set new records of intensity and strength. The coastal lands will be wracked with violence. There will be much destruction of land before the turning of the axis.

In your country (America), the areas which will suffer the most will be the West coast as it disappears, and the East coast as it also takes its turn at being destroyed. New York will not stand again as the giant trade center. Japan will lose its power center and they will be caring for their own and will not be able to take advantage of the world situation for gain. The trading centers of the world will be destroyed by the planet herself. All the brains in the world will not be able to stop the onslaught of the Earth's rages.

The polar ice caps will begin to melt. As they split apart, large chunks of ice will break off and become a danger to ships and to seashores. There will be concern that the seas will flood many land masses. This will begin slowly and be unpredictable. The melting polar caps will cause the water level of the oceans to rise, changing the seacoast around the world forever.

The state and city of New York will begin to be flooded with swollen seashores. The waters along this coast will rise and much of New York City will have to be evacuated. The evacuation will be permanent, but at the beginning many will think it is temporary. The polar ice caps will have begun to melt, and that will be cited as the reason for the flooding. In reality, the earth will have already begun to sink in preparation for the major event. Along this coast there will be earthquakes and unusual electrical storms. The storms will intensify and add to the problem of the flooding. The gulf will rise, and the state of Florida will lose much of its land back to the sea. The Indies will be besieged by storms, but will not evacuate its population. The middle part of Russia will have floods and rains which will be unequaled in their damage. This isolated part of the world will be in the news and become known worldwide.

The ocean currents will be changing, and the magnetic fields of the Earth will be moved about. Barometric and magnetic fields will set new records. The core of the earth will spin and churn. The animals on the land and in the sea will be dying off in great numbers, and many wild animals will become extinct.

The droughts which are affecting the world will continue. Those who reside on the African continent will be most affected, but other parts of the world will also begin to feel the droughts. There will be famines in those lands which have had the most wars. World hunger will be increasing; worldwide damaging winds will occur in many lands.

The winter months will be colder and in many areas wetter. Much snow will accumulate in places which have never had snow before. There will also be floods, mudslides, and torrential rains likely in all parts of the globe. These changes will be caused by the shifting molten masses in the interior of the Earth. Inside the Earth the changes are already occurring. The oceans will churn with out-of-season storms. There will be hurricanes, tornadoes, tidal waves of huge magnitude. Weather patterns will change dramatically all over the world. It will seem as if the heavens have opened an avalanche of storms. The winds will become unsettled and unpredictable. There will be tornadoes in areas where there have never been tornadoes. Hurricanes will come out-of-season. All these changes will send an alarm throughout the world. During these years there will be changes in the atmosphere as it begins to act different. There will be so many changes in the weather and oceans that the government officials will have to notice.

In the news there will begin to be predictions about the coming events. There will be many who will still doubt. But some of the scientists will come out in favor of having research to study the changing patterns around the globe. The scientists will bring gathered data out into the open. At this time, your government leaders around the world will become concerned. The world leaders will be in an uproar and will begin consulting amongst each other. Countries which have been hostile to each other will begin dialogue on this all important topic. There will be speculation about the end of the world. This will only be the beginning of the end of this era.

There will be increases in the frequency and activity of earthquakes and volcanoes, many in areas where earthquakes have never happened before. The African continent will find its long lost volcanoes which have been extinct for centuries. The earthquakes will begin on the Eastern seaboard of America. There will be small tremors in the northeast of your country. Earthquakes will be felt in Japan and India. This will be unusual, for in India there have not been earthquakes of this size for centuries. Many earthquakes will happen in Japan in varying intensity. South Africa will begin to feel the tremors of the Earth as it is off-centered. The Pacific ocean will begin not to be peaceful as its name implies. The California coast will be wracked by giant earthquakes. A giant earthquake will happen in California and the sea will recover that part of the country. These giant quakes will also hit in other parts of the world, such as Italy, Greece, Russia, Turkey, China, Colombia and in the Himalayan mountains. The earthquakes which have come before will seem small in comparison with these giant monsters. There will be tremors felt all around the world. It will seem that the whole planet is undergoing an earthquake at the same time. These tremors will cause much damage around the world. You will still have worldwide news coverage, so you will hear about other parts of the world. Some areas will sustain much damage and death. Not one nation will stand untouched. Much planning and many preparations for the future earthquakes will begin.

The mid-Pacific islands will begin to grow as volcanoes under the sea spout forth their lava. The South Seas Islands will grow the most. Those islands which have the large statues will come back to their original size.

These are truly the final years, as you see, for the next big happening will be the turning of the planet Earth onto its side, as it juggles and shakes the oceans out of their beds and new ground is brought up from the depths of the oceans. The new lands will appear in these last years. Atlantis will rise and become known.

There were two great civilizations who reached your era of technology in antiquity, Atlantis and Pacifica (Lemuria). This civilization was called Pacifica, as it was truly a peaceful place, but you know it as Lemuria. These civilizations also had to endure these tragedies, as you on earth will. This is not the first time this era of technology has been reached. It is not the only time the planet has been abused. In these old eras, the people also became puffed up with their own importance. They also thought their technology had all the answers.

The truth is that the very planet will flip over on its side. All the land that has lain fallow over these last millions of years will take its place above the seas. This will be new land, and the stars and planets will be stationed differently.

The turning of the axis is the story of the book of Revelation. It is the story told in all cultures and in all races. It has been told throughout eternity, for this is a major event. All the cultures of the world have had in their antiquity, stories of these last times, the last times of this era, and not of the world. Nor is this the end of the world. Make sure you understand that this is not the end, but the beginning of a new era and a new understanding. The need to prepare is now, right before the birth of this new era.

Japan will fall into the sea and that part of the world will become frozen wasteland. China will have much land laid to rest under mountains of ice. The continent of Europe will almost vanish as it returns to the sea. The Middle East will change climate and become quite cold and mountainous. The deserts will bloom and become virtually gardens of green growing specimens. Much of the land which your country (America) occupies will be turned on its side and will all become a warm climate. It will be quite different than it is now. Yes, I know you have hot summers and cold winters. Your climate will become milder and it will not be too hot or too cold.

After the turning of the Earth, there will be two suns. This will become a binary solar system. A new sun will enter the immediate solar system.

The Northern Lights will be seen further south than usual. The sights and signs from the sky will include this phenomenon. There will also be reports of unusual sun activity. The stars will be giving off different beams, which will be picked up by your scientists. There will be unusual meteor showers, brilliant lights coming from outer space which cannot be identified by your learned ones.

There will be unusual sun and sunspot activity. There will be unusual lights coming from space, like none which have ever been seen on earth before. Debris from space will land on Earth causing craters and changes. The sky in the last five years will be very active and there will be many new stars discovered and seen by the naked eye. There will be comets which will come through your solar system. These will be new comets and some very old comets which have not been in this part of the universe in millions of years. Mary will increase her apparitions throughout the world.

There will be increased UFO activity. Those civilizations which are on other planets will be appearing in record numbers. More and more people will see them, and photographs will be taken by high-ranking officials. UFOs will be seen almost daily. They will come in great numbers and will try to make your governments on earth understand that they come in peace. It will be in these years that some of the aliens will suddenly appear to your world leaders and offer help. It will be as if the Angel of the Lord had come to help. But surely you who have the knowledge beforehand will see the workings of the whole Universe to save mankind.

With a new attitude, your world leaders will be ready to hear of ways to save the populations and of the coming turning of the Earth. The extraterrestrials, as you call them, will be of much help, but still your leaders will not want to give any hint that these negotiations are transpiring.

They will set up stations in areas of the world which are not as inhabited. There are but a few of these areas left, but high in the mountains and in some of the desert areas they will set up substations. On other planets which are near Earth, there are already stations which have been prepared and are being stocked for man's arrival. In the last days, these will hold many people for survival.

There will not be a third world war. The countries of the world will be too concerned with sustaining themselves against the elements to wage war.

Through this whole sad scene will come hope, and the people of the nations will begin to turn to their individual concepts of God. Each man, woman, and child will begin to seek his maker. The attitude of the collective consciousness will have changed, and it will be an attitude of humility and fear.

This time I have been given permission to warn you ahead of time. It will be by the prayers and meditations of a few that many will be saved. It will be by those who earnestly pray and meditate for all mankind that the technology and advancements of this era will not be lost for all time.

The people of the world will be in a mood of humbleness and genuine seeking of God. This is good, for through last minute seeking, many will be spared their spiritual lives and advancement.

God will help anyone who truly and sincerely seeks Him. He does not care that you have wasted, perhaps, the whole lifetime on earth. It's the sincere seeking of your soul which makes God happy.

I come to tell you beforehand, seek God and look for the answers to your spiritual questions ahead of time, for in these last few years you will be too busy maintaining your sanity and your individual lives. In these last years, those who are sincerely seeking God will have the advantage. The Divine will become your lighthouse and your lamp. The Divine within will lead you to safe areas and to those places which will save your physical lives. The Divine will not be anyplace except in your own mind. The door will be in your heart. Only within can you find all the help you desire.

Through all the disasters, commerce and trade will continue among the nations. The leaders of the world will band together and seek answers to questions which should have been handled in past years, but the cooperation and the attitude of togetherness will sustain the world.

Message # 9

The Grail Message
THE GREAT COMET

For years now, knowing ones have been speaking of the coming of this especially significant STAR. The number of those who await it is continually increasing, and the indications become more and more definite, so much so in fact it is to be expected soon. But what it really signifies, what it brings, has not yet been rightly explained. It is thought that it brings upheavals of an incisive nature. But this star portends more.

It can be called the Star of Bethlehem because it is of exactly the same nature as that was. Its power sucks the waters up high, brings weather catastrophes and still more. When encircled by its rays, the earth quakes.

Unerringly and unswervingly, the COMET pursues its course, and will appear on the scene at the right hour, as already ordained thousands of years ago.

The first direct effects have already begun in recent years. For anyone who wishes neither to see nor to hear this, and who does not perceive how ridiculous it is still to maintain that all the extraordinary things which have already happened are of everyday occurrence, there is naturally no help for you. He either wishes to act like an ostrich out of fear, or he is burdened with an extremely limited understanding. Both types must be allowed to go serenely on their way; one can only smile at their easily refutable assertions.

Message # 10

A TWELFTH CENTURY PROPHECY ABOUT AMERICA

One of the earliest visions ever made concerning the future of the United States was made by **Saint Hildegard**, three centuries before the New World was discovered. She predicted that one day there would come forth "a great nation across the ocean that will be inhabited by peoples of different tribes and descent" — a good description of the American "melting pot" of immigrants from many foreign countries. For this future nation, however, the Saint sounded several warnings, all of which would come about approximately at the time of the appearance of a "great comet." "Just before the comet comes," Saint Hildegard forecast, "many nations," including America, "will be scourged by want and famine." When the comet does finally pass over... "The great nation will be devastated by earthquakes, storms, and great waves of water, causing much want and plagues. "The Comet, by its tremendous pressure, will force much out of the ocean and flood many countries... All coastal cities will live in fear, with many destroyed..."

Message # 11

Johannes Friede (1204-1257) was an eminent thirteenth-century Austrian monk. Here are his words;

70

When the great time will come, in which mankind will face its last, hard trial, it will be foreshadowed by striking changes in nature. The alteration between cold and heat will become more intensive; storms will have more catastrophic effects; earthquakes will destroy greater regions and the seas will overflow many lowlands. Not all of it will be the result of natural causes, but mankind will penetrate into the bowels of the earth and will reach into the clouds, gambling with its own existence. Before the powers of destruction will succeed in their design, the universe will be thrown into disorder, and the age of iron will plunge into nothingness.

When nights will be filled with more intensive cold and days with heat, a new life will begin in nature. The heat means radiation from the earth, the cold the waning light of the sun. Only a few years more, and you will become aware that sunlight has grown perceptibly weaker. When even your artificial light will cease to give service, the great event in the heavens will near...

Message # 12

The Coming Chastisement by Yves Dupont

This is from Catholic Prophecy. Let us recall briefly the situation which afflicted the Egyptians, the crossing dry-shod of the Red Sea and the prolonged duration of the day. In Mexico, on the other hand, a prolonged night was recorded as evidenced by archaeological discoveries. The passage of the comet at that time was recorded, not only in the Book of Exodus, but also in other documents: the Egyptian papyrus, a Mexican manuscript, a Finnish narration, and many more...

Will the comet to come be the same as that of Exodus? It is not impossible when we consider the description of the plagues as given in Exodus and those described in our Christian prophecies. When the tail of the Exodus comet crossed the path of the earth, a red dust, impalpable, like fine flour, began to fall. It was too fine to be seen... but it colored everything red, and the water of the Egyptians was changed into blood color... After the fine rusty pigment fell over Egypt, there followed a coarser dust—"like ash," this is recorded in Exodus, for then it was visible...

The narrative of the Book of Exodus confirms this and is in turn corroborated by various documents found in Mexico, Finland, Siberia, and India. It is therefore certain that a comet crossed the path of the earth more than 3000 years ago, causing widespread destruction. This is the kind of phenomenon which is soon to strike the earth again.

Message # 13

Aron Abrahamsen

In regards to Earth Changes expect the following: First there will be a severe earthquake followed by a tidal waves near the east coast of India. This will cause much damage along the southern tip all the way to the northern part. The coastline will be devastated up to 45-65 miles inland. Much of that land will be under water, some parts will go as far as 75 miles inland and remain there. The tidal wave at its peak will be 235 feet high. It will be as a wall of water descending upon the land. The wall of water will be very, very wide, starting with the southern tip of India. A few days later another wave will hit the mid-section of India. Then a third wave will hit the northern area several days after the second wave. Each time the wave will increase its intensity and force. The first wave will be about 125 feet high, the second one will be 50 feet higher, and the third one will be something like 235 feet high.

Approximately 12 days after the third wave, an earthquake will manifest itself at the bottom of the ocean adjacent to Japan. This will also affect the Philippine Islands in the deepest part of the ocean around these islands. This will be a very deep quake, originating many miles below the ocean floor. This will cause another tidal wave which will devastate the Philippine Islands as well as Japan. The people will not have much warning of this quake. There will only be this one with a magnitude of 9.5. It will devastate Japan, demolish Tokyo and Kyoto.

Three weeks after these events, there will be a quake below the ocean floor around Hawaii. The first one will be near the island of Hawaii. There will be a rift going in the east-west direction under the islands. This quake will come with such great force that it, and the accompanying tidal wave, will cause great devastation.

Four weeks later another quake below the sea bottom will occur near California resulting in a rift which will cause lands to open up north to south along the San Andreas fault. The San Joaquin and Sacramento valleys will sustain great damage. Following this, an opening of a rift between Seattle and Tacoma, in the state of Washington, going east-west almost into Idaho. This rift will be wide enough for water to find its way in there.

Two months later after the beginning of the rift in California, there will be an earthquake in the Palmdale and San Diego areas. It is difficult to say that these two will happen simultaneously, but it will appear to be like one in both places for the timing will be so close. But they are actually two separate quakes. The point of impact (epicenter) in Palmdale will spread out to Palm Springs and beyond. In San Diego the quake will go north and east resulting in ground breaking apart, making San Diego, for the moment an island. The quake will creep south into Baja California area causing a rift from the Pacific coast going very close to the border between these two countries [Mexico and the USA]. Eventually that rift will open up and water will fill it. Along the west coast of the USA land will begin to fall apart, going from Los Angeles into San Bernadino. From Santa Barbara to San Luis Obispo will be an island extending into the city of Modesto. This island will be called the Greater Santa Barbara Island.

The effect of the earth changes, together with the disruptive forces form Lake Erie and the St. Lawrence Seaway, will be very surprising, especially for the states of New York, Pennsylvania, and Ohio. Expect large inundations, not just one or two feet, but extending for ten or fifteen miles form the eastern-most part of Ohio, bordering Lake Erie, to the western-most part along the lake front. It will be severe, stretching from east to west for many miles.

This is because there is an underground fault in Lake Erie, as well as in the St. Lawrence Seaway. These faults will be activated as a result of undersea volcanic activity, causing a seismic wave which begins a tsunami right in the Seaway, pushing its way to the shoreline of Ohio, Michigan, and the Canadian shoreline. It will be devastating in all these directions, resulting in a largely widened Seaway.

Pennsylvania and New York will also be affected, those areas bordering the Lake and the Seaway. The inundation in New York and Pennsylvania will be from five to twelve miles.

There are fault lines going east-west and northwest and southeast, sort of criss-crossing Ohio. These faults have been known for some time, but not much has been said about them. There will be an activation of a fault in the mid portions of Ohio from Columbus east-west to the borders. The magnitude will be in the region of 6. This will have some severe effects. It will be more quiet going farther south.

Message # 14

Mother Shipton was first published in the 17th century England. Apparently she knew how, why, when, and what time she was going to die. She saw what modern man was going to be up to as clearly as if watching TV. Here is but a sample of what she wrote.

A fiery Dragon will cross the sky
Six times before this earth shall die
Mankind will tremble and frightened be
For the sixth heralds in this prophecy.

For those who live the century through
In fear and trembling this shall do.
Flee to the mountains and the dens

The bog and forest and wild fens.

For storms shall rage and oceans roar
When Gabriel stands on sea and shore
And as he blows his wondrous horn
Old worlds die and new be born.

There'll be a sign for all to see
Be sure that it will certain be.
Then love shall die and marriage cease
And nations wane as babes decrease.

And wives shall fondle cats and dogs
And men live much the same as hogs.
Yet greater signs there be to see
As man nears latter century

Three sleeping mountains gather breath
And spew out mud, and ice and death.
And earthquakes swallow town and town
In lands as yet to me unknown

Not every soul on Earth will die
As the Dragon's tail goes sweeping by.
Not every land on Earth will sink
But these will wallow in stench and stink
Of rotting bodies of beast and man
Of vegetation crisped on land.

But the land that rises from the sea
Will be dry and clean and soft and free
Of mankind's dirt and therefore be
The source of man's new dynasty.

And those that live will ever fear
The Dragon's tail for many year
But time erases memory
You think it strange. But it will be

And when the Dragon's tail is gone
Man forgets, and smiles, and carries on
To apply himself — too late, too late

For mankind has earned deserved fate.

His masked smile, his fate grandeur
Will serve the Gods their anger stir.
And they will send the Dragon back
To light the sky—his tail will crack

Upon the Earth and rend the Earth
And man shall flee, King, Lord and serf.
For seven days and seven nights
Man will watch this awesome sight.

The tides will rise beyond their ken
To bite away the shores, and then
The mountains will begin to roar
And earthquakes split the plain and shore.

And flooding waters, rushing in
Will flood the lands with such a din
That mankind cowers in muddy fen
And snarls about his fellow men.

Man flees in terror from the floods
And kills, and rapes and lies in blood
And spilling blood by mankind's hands
Will stain and bitter many lands

He bares his teeth and fights and kills
And secrets food in secret hills
And ugly in his fear, he lies
To kill marauders, thieves and spies

But slowly they are routed out
To seek diminishing water spout
And men will die of thirst before
The oceans rise to mount the shore.
And lands will crack and rend anew
You think it strange. It will come true.

And in some far-off distant land,
Some men — oh, such a tiny band,

Will have to leave their solid mount,
And span the earth, those few to count,

Who survive this and then
Begin the human race again.
But not on land already there
But on ocean beds, stark, dry and bare.

And before the race is built anew,
A silver serpent comes to view
And spew out men of like unknown
To mingle with the earth now grown

Cold from its heat, and these men can
Enlighten the minds of future man
To intermingle and show them how
To live and love and thus endow

The children with the second sight.
A natural thing so that they might
Grow graceful, humble, and when they do,
The Golden Age will start anew.

The Dragon's tail is but a sign,
For mankind's fall and man's decline.
And before this prophecy is done,
I shall be burned at the stake, at one,

My body singed and my soul set free.
You think I utter blasphemy, You're wrong.
These things have come to me,
The prophecy will come to be.

Message # 15

Joe Brandt

   Whether it is true or not, this story is another spiritual message.   Depending on
your psychological make-up, the repetitious statements from the others may have

left you somewhat ambivalent. Occasionally the best wake-up call for some is a simple personal story.

In 1937, a 17-year-old boy experienced a spontaneous vision while recovering from a brain concussion in a Fresno hospital in California. This boy put his entire vision down on paper by his own hand. Young Joe Brandt wrote the following while passing in and out of consciousness.

I woke up in the hospital room with a terrific headache—as if the whole world was revolving inside my brain. I remember, vaguely, the fall from my horse—Blackie. As I lay there, pictures began to form in my mind—pictures that stood still. I seemed to be in another world. Whether it was the future, or it was some ancient land, I could not say. Then slowly, like the silver screen of the "talkies," but with color and smell and sound, I seemed to find myself in Los Angeles—but I swear it was much bigger, and buses and odd-shaped cars crowded the city streets.

I thought about Hollywood Boulevard, and I found myself there. Whether this is true, I do not know, but there were a lot of guys about my age with beards and wearing, some of them, earrings. All the girls, some of them keen-o, wore real short skirts... and they slouched along—moving like a dance. Yet they seemed familiar. I wondered if I could talk to them, and I said, "Hello," but they didn't see or hear me. I guess it is something you have to learn. I couldn't do it.

I noticed there was a quietness about the air, a kind of stillness. Something else was missing, something that should be there. At first, I couldn't figure it out; I didn't know what it was—then did. There were no birds. I listened. I walked two blocks north of the Boulevard—all houses—no birds. I wondered what had happened to them. Had they gone away? Where? Again, I could hear the stillness. Then I knew something was going to happen. I wondered what year it was. It certainly was not 1937. I saw a newspaper on the corner with a picture of the President. It surely wasn't Mr. Roosevelt. He was bigger, heavier, big ears. If it wasn't 1937, I wondered what year it was... My eyes weren't working right. Someone was coming—someone in 1937—it was that darned, fat nurse ready to take my temperature. I woke up. Crazy dream.

[Then next day]. Gosh, my headache is worse. It is a wonder I didn't get killed on that horse. I've had another crazy dream, back in Hollywood. Those people. Why do they dress like that, I wonder? Funny glow about them. It is a shine around their heads—something shining. I remember it now. I found myself back on the Boulevard. I was waiting for something to happen, and I was going to be there. I looked up at the clock down by that big theater. It was ten minutes to four. Something big was going to happen.

I wondered if I went into a movie (since nobody could see me) if I'd like it. Some cardboard blond was draped over the marquee with her legs six feet long. I started to go in, but it wasn't inside. I was waiting for something to happen outside. I walked down the street. In the concrete they have names of stars. I just recognized a few of them. The other names I had never heard. I was getting bored; I wanted to get back to the hospital in Fresno, and I wanted to stay there on the Boulevard, even if nobody could see me. Those crazy kids. Why are they dressed like that? Maybe it is some big Halloween doings, but it don't seem like Halloween doings. More like early spring. There was that sound again, that lack of sound. Stillness, stillness, stillness. Don't these people know that the birds have gone somewhere? The quiet is getting bigger and bigger. I know it is going to happen. Something is going to happen. It is happening now! It sure did. She woke me up, grinning and smiling , that fat one again.

"It's time for your milk, kiddo," she says. Gosh, old women of thirty acting like the cat's pajamas. Next time maybe she'll bring hot chocolate.

Where have I been? Where haven't I been? I've been to the ends of the earth and back. I've been to the end of the world—there isn't anything left. Not even Fresno, even though I'm lying here right this minute. If only my eyes would get a little clearer so I can write all this down. Nobody will believe me, anyway. I'm going back to that last moment on the Boulevard. Some sweet kid went past, dragging little boys (twins, I guess) by each hand. Her skirt was up—well, pretty high—and she had a tired look. I thought for a minute I could ask her about the birds, what had happened to them, and then I remembered she hadn't seen me. Her hair was all frowzy, way out all over her head. A lot of them looked like that, but she looked so tired and like she was sorry about something. I guess she was sorry before it happened—because it surely did happen. There was a funny smell. I don't know where it came from. I didn't like it. A smell like sulphur, sulfuric acid, a smell like death. For a minute I thought I was back in chem. [chemistry].

When I looked around for the girl, she was gone. I wanted to find her for some reason. It was as if I knew something was going to happen and I could stay with her, help her. She was gone, and I walked half a block, then I saw the clock again. My eyes seemed glued to that clock. I couldn't move. I just waited. It was five minutes to four on a sunny afternoon. I thought I would stand there looking at the clock forever waiting for something to come. Then, when it came, it was nothing. It wasn't nearly as hard as the earthquake we had two years ago. The ground shook, just an instant. People looked at each other, surprised. Then they laughed. I laughed, too. So this was what I had been waiting for? This funny little shake? It meant nothing.

I was relieved and I was disappointed. What had I been waiting for? I started back up the Boulevard, moving my legs like those kids. How do they do it? I never found out. I felt as if the ground wasn't solid under me. I knew I was dreaming, and yet I wasn't dreaming. There was that smell again, coming like from the ocean. I was getting to the 5 and 10 store and I saw the look on the kids faces. Two of them were right in front of me, coming my way.

"Let's get out of this place. Let's go back East." He seemed scared. It wasn't as if the sidewalks were trembling—but you couldn't seem to see them. Not with your eyes you couldn't. An old lady had a dog, a little white dog, and she stopped and looked scared, and grabbed him in her arms and said: "Let's go home, Frou-Frou. Mama is going to take you home." That poor lady, hanging on to her dog.

I got scared. Real scared. I remembered the girl. She was way down the block, probably. I started to run. I ran and ran, and the ground kept trembling. I couldn't see it. But I knew it was trembling. Everybody looked scared. They looked terrible. One young lady just sat down on the sidewalk all doubled up. She kept saying, "earthquake, it's the earthquake," over and over. But I couldn't see that anything was different.

Then, when it came, how it came! Like nothing in God's world. Like nothing. It was like the scream of a siren, long and low, or the scream of a woman I heard having a baby when I was a kid. It was awful. It was as if something—some monster—was pushing up the sidewalks. You felt it long before you saw it, as if the sidewalks wouldn't hold you anymore. I looked out at the cars. They were honking, but not scared. They just kept moving. They didn't seem to know yet that anything was happening. Then, that white car, that baby half-sized one came sprawling from the inside lane right against the curb. The girl who was driving just sat there. She sat there with her eyes staring, as if she couldn't' move, but I could hear her. She made funny noises.

I watched her, thinking of the other girl. I said that it was a dream, and I would wake up. But I didn't wake up. The shaking had started again, but this time different. It was a nice shaking, like a cradle being rocked for a minute, and then I saw the middle of the Boulevard seemed to be breaking in two. The concrete looked as if it were being pushed straight up by some giant shovel. It was breaking in two. That is why the girl's car went out of control. And then a loud sound again, like I've never heard before—then hundreds of sounds—all kinds of sounds: children, and women, and those crazy guys with earrings. They were all moving, some of them above the sidewalk. I can't describe it. They were lifted up...

And the waters kept oozing—oozing. The cries. God, it was awful. I woke up. I never want to have that dream again.

It came again. Like the first time, which was a preview, and all I could remember was that it was the end of the world. I was right back there—all that crying. Right in the middle of it. My eardrums felt as if they were going to burst. Noise everywhere. People falling down, some of them hurt badly. Pieces of buildings, chips, flying in the air. One hit me hard on the side of the face, but I didn't seem to feel it. I wanted to wake up, to get away from this place. It had been fun in the beginning, the first dream, when I kind of knew I was going to dream the end of the world or something. This was terrible. There were older people in cars. Most of the kids were on the street. But those old guys were yelling bloody murder, as if anybody could help them. Nobody could help anybody. It was then I felt myself lifted up. Maybe I had died. I don't know. But I was over the city. It was tilting toward the ocean—like tilting a picnic table.

The buildings were holding, better than you could believe. They were holding. They were holding.

The people saw they were holding, and they tried to cling to them or get inside. It was fantastic. Like a building had a will of its own. Everything else breaking around them, and they were holding, holding. I was up over them—looking down. I started to root for them. "Hold that line," I said. "Hold that line. Hold that line. Hold that line." I wanted to cheer, to shout, to scream. If the buildings held, those buildings on the Boulevard, maybe the girl—the girl with the two kids—maybe she could get inside. It looked that way for a long time, maybe three minutes, and the three minutes was like forever. You knew they were going to hold, even if the waters kept coming up. Only they didn't.

I've never imagined what it would be like for a building to die. A building dies just like a person. It gives way, some of the bigger ones did just that. They began to crumble, like an old man with palsy, who couldn't take it anymore. They crumbled right down to nothing. And the little ones screamed like mad—over and above the roar of the people. They were mad about dying. But buildings die.

I couldn't look anymore at the people. I kept wanting to get higher. Then I seemed to be out of it all, but I could see. I seemed to be up on Big Bear near San Bernardino, but the funny thing was that I could see everywhere. I knew what was happening.

The earth seemed to start to tremble again. I could feel it even though I was up high. This time it lasted maybe twelve seconds, and it was gentle. You couldn't believe anything so gentle could cause so much damage. But then I saw the streets of Los Angeles—and everything between the San Bernardino mountains and Los Angeles. It was still tilting towards the ocean, houses, everything that was left. I could see the big lanes—dozens of big lanes still loaded with cars sliding the same way. Now the ocean was coming in, moving like a huge snake across the land. I wondered how long it was, and I could see the clock, even though I wasn't there on the Boulevard. It was 4:29. It had been half an hour. I was glad I couldn't hear the crying anymore. But I could see everything. I could see everything.

Then, like looking at a huge map of the world, I could see what was happening on the land and with the people. San Francisco was feeling it, but she was not in any way like Hollywood or Los Angeles. It was moving just like that earthquake movie with Jeanette McDonald and Gable. I could see all those mountains coming together... I knew it was going to happen to San Francisco—it was going to turn over—it would turn upside down. It went quickly, because of the twisting, I guess. It seemed much faster than Hollywood, but then I wasn't exactly there. I was a long way off. I was a long, long way off. I shut my eyes for a long time—I guess ten minutes—and when I opened them I saw Grand Canyon.

When I looked at Grand Canyon, that great big gap was closing in, and Boulder Dam was being pushed from underneath. And then, Nevada, and on up to Reno. Way down south, way down. Baja California. Mexico too. It looked like some volcano down there was erupting, along with everything else. I saw the map of South America, especially Colombia. Another volcano—eruption—shaking violently. I seemed to be seeing a movie of three months before—before the Hollywood earthquake. Venezuela seemed to be having some kind of volcanic activity. Away off in the distance, I could see Japan, on a fault, too. It was so far off—not easy to see because I was still on Big Bear Mountain, but it started to go into the sea. I couldn't hear screaming, but I could see the surprised look on their faces. They looked so surprised. Japanese girls are made well, supple, easy, muscles that move well. Pretty, too. But they were all like dolls. It was so far away I could hardly see it. In a minute or two it seemed over. Everybody was gone. There was nobody left.

I didn't know time now. I couldn't see a clock. I tried to see the island of Hawaii. I could see huge tidal waves beating against it. The people on the streets were getting wet, and they were scared. But I didn't see anybody go into the sea.

I seemed way around the globe. More flooding. Is the world going to be drenched? Constantinople. Black Sea rising. Suez Canal, for some reason seemed to be drying up. Sicily—she doesn't hold. I could see a map. Mt Etna. Mt Etna is shaking. A lot of area seemed to go, but it seemed to be earlier or later. I wasn't sure of time, now.

England—huge floods—but no tidal waves. Water, water everywhere, but no one was going into the sea. People were frightened and crying. Some places they fell to the streets on their knees and started to pray for the world. I didn't know the English were emotional. Ireland, Scotland—all kinds of churches were crowded—it seemed night and day. People were carrying candles and everybody was crying for California, Nevada, parts of Colorado—maybe all of it, even Utah. Everybody was crying-most of them didn't even know anybody in California, Nevada, Utah, but they were crying as if they were blood kin. Like one family. Like it happened to them.

New York was coming into view—she was still there, nothing had happened, yet water level was way up. Here, things were different. People were running in the streets yelling—"end of the world." Kids ran into restaurants and ate everything in sight. I saw a shoe store with all the shoes gone in a few minutes; power might be shut off. They must control themselves. Five girls were running like mad toward the YMCA, that place on Lexington or somewhere. But nothing was happening in New York. I saw an old lady with garbage cans filling them with water. Everybody seemed scared to death. Some people looked dazed. The streets seemed filled with loudspeakers. It wasn't daylight. It was night.

I saw a lot of places that seemed safe, and people were not so scared. Especially the rural areas. Here everything was almost as if nothing had happened. People seemed headed to these places, some on foot, some in cars that still had fuel. I heard—or somehow I knew—that somewhere in the Atlantic land had come up. A lot of land. I was getting awfully tired. I wanted to wake up. I wanted to go back to the girl—to know where she was—she and those two kids. I found myself back in Hollywood—and it was still 4:29. I wasn't up on Big Bear at all; I was perched over Hollywood. I was just there. It seemed perfectly natural in my dream.

I could hear now. I could hear, someplace, a radio station blasting out—telling people not to panic. They were dying in the streets. There were picture stations with movies—some right in Hollywood—these were carrying on with all the shaking. One fellow in the picture station was a little short guy who should have been scared to death. But he wasn't. He kept shouting and reading instructions. Something about helicopters or planes would go over—some kind of planes—but I knew they couldn't. Things were happening in the atmosphere. The waves were rushing up now. Waves. Such waves. Nightmare waves were rushing up now. Waves. Such waves. Nightmare waves.

Then, I saw again Boulder Dam going down—pushing together, pushing together breaking apart—no, Grand Canyon was pushing together, and Boulder Dam was breaking apart. It was still daylight. All these radio stations went off at the same time—Boulder Dam had broken.

I wondered how everybody would know about it—people back East. That was when I saw the "ham radio operators." I saw them in the darndest places, as if I were right there with them. Like the little guy with glasses, they kept sounding the alarm. One kept saying: "This is California. We are going into the sea. This is California. We are going into the sea. Get to high places. Get to the mountains. All states west—this is California. We are going into the... we are going into the..." I thought he was going to say "sea," but I could see him. He was inland, but the waters had come in. His hand was still clinging to the table, he was trying to get up, so that once again he could say: "This is California. We are going into the sea. This is California. We are going into the sea."

I seemed to hear this, over and over, for what seemed hours—just those words—they kept it up until the last minute—all of them calling out, "Get to the mountains—this is California. We are going into the sea."

I woke up. It didn't seem as if I had been dreaming. I have never been so tired. For a minute or two, I thought it had happened. I wondered about two things. I hadn't seen what happened to Fresno and I hadn't found out what happened to that girl.

I've been thinking about it all morning. I'm going home tomorrow. It was just a dream. It was nothing more. Nobody in the future on Hollywood Boulevard is going to be wearing earrings—and those beards. Nothing like that is ever going to happen. That girl was so real to me—that girl with those kids. It won't ever happen—but if it did, how could I tell her (maybe she isn't even born yet) to move away from California when she has her twins—and she can't be on the Boulevard that day. She was so gosh-darned real.

The other thing—those ham operators—hanging on like that—over and over—saying the same thing:

"This is California. We are going into the sea. This is California. We are going into the sea. Get to the mountains. Get to the hilltops. California, Nevada, Colorado, Arizona, Utah. This is California. We are going into the sea." I guess I'll hear that for days.

[For some time it has been known scientifically that a good portion of the coast of California is just a shell that projects out over the water like a shelf. It is known that one can go out from the shoreline a short distance on the California coast and then it just drops off...there seeming to be no bottom. The largest part of California is washed out underneath. There are only upright supports holding the coastline of California. Scientists inform us that the coastline of California is moving north-west at the rate of two inches a year. This places a great strain on the San Andreas Fault which extends from lower California to Palm Springs, near San Bernardino, then over to Palmdale and extending up to the San Francisco Bay area and to Point Arena. All this land west of the fault is moving every year. The stanchions or supports cannot move, as they are part of the ocean bed, making it necessary for them to incline or tip.

One day these supports will tip enough that they cannot bear the weight of the land. This will be at the time of a Great Earthquake. The west side of the San Andreas fault will break off and slide into the sea...

Message # 16

Nancy Lieder is either an amazing modern day intuitive, or where she says her information is coming from is true; either way it's for you to decide. No matter how prophecy arrives, it stands on its own merits. **Zetatalk** was launched on the internet in 1995. It quickly became very popular worldwide and has been translated in at least 13 languages.

Verification and scientific proof are being added on a regular basis to a complement internet site of 3000 plus pages called Troubled Times as the prophecy is fulfilled again and again. This scientific documentation on the Troubled Times website is supplied by people from terrestrial sources. The prophecies from the other site entitled Zetatalk have been very accurate thusfar and are being continually corroborated by Troubled Times. I've studied many intuitives over the years, and I've rarely run across this level of meticulous scrutiny to authenticate anyone's prophetic utterances with hard science.

Although the leaders of the world are fully aware of the situation, announcements will not be forthcoming. Several grass-roots sources such as Zetatalk and Troubled Times are helping lead the way.

On June 30, 2000 the Zetatalk Accuracy TOPIC in Troubled Times was updated with a massive amount of substantiation on wide ranging topics, including the accuracy of Zetatalk predictions on <u>Cataclysm Masks</u>, <u>Weather Changes</u>, <u>Migrations</u>, <u>Domino Quakes</u>, and <u>Summer Snowstorms</u>. Within 24 hours, access to Zetatalk had been blocked, for the first time, by the Emirates Internet in Saudi Arabia, which is so far the only country in the world that has banned Zetatalk. This was done July, 1st, 2000 the day after the substantiation was posted. The Emirates of Saudi Arabia is one of the most oppressive government dictatorships in the world and have decided to completely suppress one more grass-roots source from this time forward.

There is a complete record of what's been added to this site in chronological order from day one in the "What's New" section of her internet site. She has written a book called *Zetatalk*. The internet site is <u>http://www.zetatalk.com/</u>   The Zetas are said to be telepathically relaying information to Nancy.

Here are a few non-zeta discoveries I've made:

1.)   http://www.disclosureproject.org/ Apparently, many in-the-know current and previous goverment and military officers, officials and employees have been petitioning Congress to allow full-disclosure of what the military, and other branches of the government, knows about the alien issue. I don't believe this will happen. In my opinion the controller operatives in the government and Nasa won't allow it. The clarity of this whole pattern of deception, of which I continue to refer to, should be sharpening up with this in mind.

2.) Within the last two years, a high official from the Catholic Church has gone on TV in Italy 7 times to carefully inform the public that Aliens and UFOs are real and how it fits into the doctrine of the church. This was done as a preemptive strike so the church wouldn't lose more followers as governments of the world begin to be more forthcoming on the issue.

3.) http://www.ecologynews.com/cometa.html The official government of France shortly thereafter released a lengthy document outlining all they know, detailing incident after incident. The president of France and all the top military defense officials signed off on the document. This document also stated it was curious that the United States is still denying what they know. France said the US government started covering up the issue after the Roswell incident in 1947 and continues to this day. The media controllers didn't allow the Church announcement or French government report and disclosure to make it into major news outlets in the true-blue, honest, U.S. of A.

The archives of Nancy's radio shows can be listened to from the Jeff Rense radio show via the internet. Here is a minor taste of what will happen during the passage, out of the Zetatalk Pole Shift section:

When the giant comet positions itself exactly between the Earth and its Sun, things change. The Earth then has its greatest advocate for its previous alignment, the Sun and its magnetic alignment, negated. The Earth hears only the magnetic voice of the giant comet, so to speak, which stands between the Earth and its former magnetic commander, the Sun. You are aware that your Earth is heavier at its molten core, which is rumored to be composed primarily of iron. This is not entirely untrue, but regardless of the composition, the Earth's core is more sensitive to the magnetic alignment than the crust. The core grips the crust, and is not as liquid as one might think. There is friction. There is the tendency for the whole to move as one, despite their differing magnetic allegiances.

The pole shift is in fact a movement of the interior of the Earth, the core, to come into alignment with the giant comet. The 12th Planet, due to its massive size in comparison to the Earth, dominates the magnetic scene, and it is in this regard that gravity comes into the pole shift equation. The Earth's crust resists aligning with the giant comet, being caught in a web of magnetic pulls from its immediate neighborhood. In other words, the Earth's crust wants to stay with the old, established, magnetic pull, while the core of the Earth, having less allegiance and attachment to the neighborhood, listens to the new voice. There is a great deal of tension that builds between the crust of the Earth and the core of the Earth. This tension is released when the core of the Earth breaks with the crust, and moves. However, the core of the Earth drags the crust with it as it turns to align anew.

The pole shift is therefore sudden, taking place in what seems to be minutes to humans involved in the drama, but which actually takes place during the better part of an hour. There are stages, between which the human spectators, in shock, are numb. At first there is a vibration of sorts, a jiggling, as the crust separates in various places from the core. Then there is a Slide, where the crust is dragged, over minutes, to a new location, along *with* the core. During the slide, tidal waves move over the Earth along the coast lines, as the water is not attached and can move independently. The water tends to stay where it is, the crust moving under it, essentially. When the core finds itself aligned, it churns about somewhat, settling, but the crust, more solid and in motion, proceeds on. This is in fact where mountain building and massive earthquakes occur, just as car crashes do their damage on the point of impact, when motion must *stop*.

Weak spots among the Earth's crustal plates give way. The Pacific Ocean will shorten, and the Atlantic widen. Subducting plates will subduct greatly. Mountain building will occur suddenly, primarily increasing in areas already undergoing mountain building. All told, the better part of an hour, but at certain stages only minutes. Plants survive, as they are rooted and their seeds are everywhere, and animals including man survive because they travel *with* the moving plates of the Earth and experience no more severe a shock when the plates stop moving than they would during a Richter 9 earthquake. Where mountain building occurs when the plates *stop* moving, the stoppage is not simply a sudden jolt, like a car hitting a brick wall. *All* is in motion, and the stoppage is more like a car hitting a barrier of sand-filled plastic barrels — a series of small jolts, occurring in quick succession.

At this time, we estimate that the giant comet will come to within 14 million miles of the Earth. The strength of its magnetic field at that distance will be such that the comet's North Pole, angled essentially in the same direction as the Earth's North Pole, forces the Earth's North Pole to evade the pressure and accommodate its larger brother by swinging south to the bulge of Brazil. This alignment will not change if the distance between the sibling planets changes, but the speed and vigor of the shift would be so affected by a closer passage. The height of tidal waves and consequent inland inundation would be so affected. The heat of land masses above subducting plates, where friction can cause the ground to melt, would be so affected. And the violence of shifting winds would certainly be so affected.

Pole Shift section. Firestorms.

The cataclysms come with geological upheavals, volcanic eruptions, some parts of the Earth heating up to fiery temperatures, and in some cases firestorms from the sky. These firestorms are caused by reactions of atmospheric gasses to the Turmoil going on. Petrocarbons are in essence created due to the flashes of lightning and intense heat due to passage over open volcanoes, and these petrocarbons rain down, afire, at times. With the atmosphere scattered, chemicals in the comet's tail similar to your petrol chemicals do not flash in a quick consummation into water and carbon dioxide, but descend close to the surface of the Earth before bursting into flame — a firestorm, killing all beneath it. All this has been reported in ancient times, as humans observed accompaniments to the cataclysms. This type of activity sets forests afire. Where vegetation regrows, from seeds and roots, many areas will nevertheless be denuded of vegetation for some time.

Pole Shift section. Tidal Waves.

As great as the danger to humans and the fauna and flora of the Earth that earthquakes bring, greater still is the devastation that the shifting waters will bring. There are several factors at play. Water is more liquid than the core of the Earth, and certainly more liquid than terra firma. Where the Earth, dragged by its core, is Shifting into a new, albeit temporary, alignment with the giant comet, its waters resist greatly. Thus the waters slosh over the nearby land, in the direction opposite to the shift. This is lessened by a tendency of the waters directly under the giant comet to rise up to meet the comet. The waters heap up, in what appear to be giant waves. This tends to lessen the sloshing over a shoreline on the comet side, but has no effect on the water's movement on the dark side of the Earth.

The Earth's record of gigantic tidal waves, which the establishment is desperate to explain in other than pole shift terms, is caused by the frequent pole shifts. This is the case even in situations where a plate adjustment affects hundreds of miles of ocean bottom, creating a massive line of compressed water which promptly moves in the only directions it can - to the right, left, and *up*.

Troubled Times is also devoted to getting out the word of Planet X and creating solutions for all the problems which can be anticipated in the aftertimes. Many years and 1000's of people have put their time in this effort. It was the Troubled Times team in Europe that used the coordinates supplied by Nancy that spotted Planet X in the first months of 2001. Switzerland, Arizona and South America were the locations of the three observatories. See "What's New" section for March and April 2001 for more information about the 3 observatories. By the time you read this spotting, Planet X will have become more common. Go to zetatalk.com for more information about future coordinates.

Most importantly is a section called "Troubled Times Zeta advice on locations." This should be studied in detail, for this will determine the survivability of your area. Equally important is the information about dome homes. Keep safe from strong earthquakes and hurricane winds inside the strongest structure known to man. In the right location a dome home could secure your loved ones' survival.

Message # 17

Lori Adaile Toye

Lori says she has been in direct communication with what she refers to as a group of ascended Masters since her childhood. She says that the Maps she has created come directly from the Masters. They are known as the "I Am America" maps. The maps detail what the world will look like after the pole shift. Her maps are very similar to Gordon Michael Scallion's except they contain 10 times the detail of Gordon's. She has one of the United States, the entire world and a six map scenario of possibilities for the USA. Lori's husband, Lenard, and she have written several prophetic books. Lori's books contain so much detail concerning what is going to happen and its spiritual significance, I hardly know where to begin. If you're so inclined to read spiritual material in regard to Earth Changes, her works may be the absolute best. I highly recommend them. An absolutely positive beautiful spiritual position that comes purely from the deep end of the pool. Your soul will become touched with emotion wrangling about in her written work. There is however some very pointed material that is meant simply to offer explanation for events but still may shock. One is entitled *"Volume Three New World Atlas."* From one of the ascended masters in that work comes these words:

The Earth's axis begins to go through its gyrations, these kinds of tectonic movements happen nearly overnight. This is not a long, drawn-out process. It is as though you wake up one morning and the sea coast is in your front yard if you're in southern England. And it is not something that is proceeded by a long-term warning either, nor will it necessarily be picked up on seismographs. It just happens.

The masters have always spoken about ice sheeting. The ice sheeting occurs as the polar caps melt, more and more water is exposed to the sun. As this happens, it creates a vaporization very much like the formation of clouds. However, the ozone that normally provides a protective barrier from the sun's rays, or a certain spectrum of light rays, now becomes an incubation area where water steams and rises into the atmosphere, then immediately freezes. In this immediate freezing, whole cloud masses that can normally cover parts of a continent, become instantaneously frozen. When this happens, the gravitational pull of the mass of frozen clouds is too much for the dynamics of the atmosphere to sustain. This frozen mass literally falls to the ground as a great continental size sheet of frozen water vapor.

When you will take an air flight, you see hundreds of square miles of large clouds. If you can imagine them instantaneously frozen and then for them to just drop as though they were ice cubes from an ice maker, you can get a visualization of ice sheeting.

These sheets of ice will compress and compact as thick as any type of cloud mass can. So if a cloud mass is as much as a mile thick, you may have a compression of an ice sheet a quarter mile thick. Imagine that you are dealing with something maybe twelve hundred feet thick and it falls upon a city.

Last Message

Mark Hazlewood

No, I'm not really any sort of intuitive or prophet. Giving you this synopsis of the few others may have been some of my best work. I've admittedly become overwhelmed with emotion writing this book at times. Sometimes knowing what is about to occur feels like a heavy weight. Finishing up this book and getting it out will lighten the burden of knowing. I am very optimistic about the future; it's just so many billion people are about to lose their lives. If I concentrate on that too much, I can't continue to write. My philosophical beliefs do see the perfection that is not unlike the laws of the jungle. With humans, the healthiest and most cooperative will survive.

Now the strongest and most ruthless I believe will also survive for a time. My philosophical beliefs see the later group being gradually segregated to a greater degree without them knowing it. I don't believe the new world is meant for their type. After a while I see them killing each other off, leaving just the cooperative sentient beings. How long this sifting out process will take I'm not sure. How I believe this will happen is not relevant for it's just a belief. My focus has been to make certain that you know that Planet X is real and passing shortly. So many other works that have this knowledge of X have muddied up the waters and filled their messages with philosophy, that the clarity of X has been lost. I give just a sprinkling of my personal philosophical beliefs without much explanation so my focus remains intact.

There are some that believe that man's current level of consciousness has turned the tide, and in some way all of this can be avoided and Planet X's orbit changed. I wish I could say that there was even a slight chance for this idea; alas not a bit of evidence for this notion exists. This is simply a fantasy of a few who are a little too dreamy-eyed in their perceptions of this world or can't face the reality of the situation we face or don't have the facts. Besides, on this note, if you are inclined to dig into the spiritual side of this event, you will find that perfection exists for this happening. The awakening that will take place after is directly related to the severity of the calamities. In other words, these events are in fact needed for the general uplifting of spirituality to transpire. An old line is appropriate here "mother nature does work in mysterious ways."

Earth change prophecies have been used for the agenda of spiritual awakening. This has been much to the good. Sometimes the vision has been given and then used to attempt to get people to change and live their lives to the highest imaginable good. With this enticement, they've said, as a result of the change of heart of humanity, the severity of the future earth traumas will be lessened. Many have bought into this story, and thus feel that we as a people have turned the tide, and whatever is coming will be minor or much delayed. Another way that this was done was to give a close but false date of occurrence for the earth changes. This date may never have been given in the first place from their vision or source. The whole point of this was to get people to change for the better, which a few individuals may have done. Take a good look at humanity today and tell me how many are living close to an ideal life? Get out of the house. Walk downtown and look around. Drive around and look at the actions and expressions on other drivers. Read the newspapers and watch the news that is spoon-fed to you and tell me if you really see a profound change having taken place. Tell me if you do see this new, magnificent, beautiful humanity. What planet are you living on? The reality is unchanged. If you prepare correctly, you have a good chance of survival; if not, see ya in the next life. Many of us have died several times before. From a very broad timeline soul perspective, it's really not a big deal. Reality is life; it goes on and on with changes in places and the bodies that are our vehicles.

Dow Corning fulfilled a contract with the US government in August, 1999, delivering 20 million body bags to California. I guess that's their way of preparing for us. Well geez, relocation, food, water, supplies, and make-shift shelters for several million people for several years just ain't in the budget. I've a dark humorous side and find this all a bit funny. It's sometimes also how I deal with the seriousness of it all. What am I supposed to do, get depressed and mad and call my Congressman?

I do believe a better humanity and mankind are on the horizon, and that will come about as a direct result of the cataclysms. The changes are needed for us to make that next evolutionary jump in our collective consciousness toward each other and our fellow beings throughout the universe. These ideas won't take hold until a few years have past after the coming pole shift. There are some that are already well ahead and think this way currently. A mega-worldwide tragedy is part of a perfect plan to get the survivors to make this leap. The leap in thinking is simple. Every action takes into account the fact that we are all connected parts of one living being. You think of others as extensions of yourself at all times. No laws, countries or religion will be needed when this realization takes hold *en masse*. The first step is recognizing the ideal.

I've had at least three lucid earth change dreams. The first puts me in New York City at the edge of Manhattan, riding in the front passenger seat of a large convertible with the top down. A turn onto one of the many streets lined with skyscrapers had just been made. Then all of a sudden the shaking started. The driver had sense enough to jam it in reverse, getting us off the street quickly that had the massive buildings side by side before all the debris rained off of them. From this scene, I was projected to the other coast where I grew up in the San Fernando Valley of Los Angeles, California.

From the San Fernando Valley, I was looking at the hills that separate North Hollywood from Hollywood. There were these curious orange gaps in this line of hills every quarter mile or so all the way to Reseda and beyond. As my curiosity peaked as to what they were, I was then projected onto one of them. The gaps were oozing lava that was flowing from the sides of these mountains. I ran down the side of this one to get away from the flowing lava.

Next I found myself deep in the valley inside this house at a party. I was explaining to a crowded group of young people, that were just enjoying themselves, that they would never be able to travel over the mountains I had just seen. I told them it would be best if they got completely out of the city, in the other direction.

Lastly, in spring of 1999, my 19-year-old son tells me of a dream he had about getting away from a tidal wave and seeing flying saucers in downtown Orlando around where we lived at the time. Within 2 days, I had a similar dream minus the saucers. In my dream, I get away from the first tidal wave only to be met by a second wave coming from the exact direction I was running to. Central Florida is one of the places that this could very well happen, being only 50 miles from either the gulf or Atlantic shores and just above sea level.

Right at this time in my life, I started getting all these indications from different people that were unconnected to each other about getting out of Florida. This went on for the next couple of weeks. I found this rather odd, because even though I knew all about the prophetic earth changes, I didn't feel a need to make plans to move until this time. After I had announced my plans to family and friends, the whole process of doing so became overwhelming and daunting. I just wanted to know why at this time I should be making these plans. I believe that you should follow your own heart when making a decision. I'm not recommending that anyone do anything other than what they are guided to do by their own intuition. Mine said leave, but I wanted a definitive reason why.

Within 24 hours, I discovered the information in regards to Planet X. Now, in my life, timing is everything. This information came to me as I was waffling about my decision to leave. This meant that this was my time to find out what was going on and why. After reading, studying and researching much material to confirm all of this until it got very repetitious, I received what I refer to as my prophetic confirming chills.

Before I move from the intuitive section, there is one more piece of information, offered from an intuitive source, to consider. This explains why X only causes a pole-shift occasionally. X's passage is always severe enough to leave its devastating mark on Earth. The last passage tree rings world-wide narrowed for several years and the sea levels dropped 15-20 feet, but there was no pole-shift. I'm sure all the volcanoes, quakes, and weather changes were horrendous, even so. The reason could have been this: That last passage may have been quicker, closer, harder, and more jerking. The core did not slip from the crust causing the mega-devastation that a pole-shift would have. The crust is not as magnetic as the core, however, during a close passage of say only 7 million miles, Earth's crust and core align to X simultaneously. It's when the core and crust don't act in unison that you get a pole-shift and very severe changes. This only happens when X passes within a certain range of miles from Earth. This coming passage of X is supposed to pass well within that range to cause a pole shift. In other words not too close so that the entire Earth aligns to X together, but close enough so the core aligns and the crust does not. This coming passage is supposed to be this worst case scenario. That is the reason why all the intuitives agree on a pole-shift this time.

My Angel of Truth

Some have no idea what a spiritual connection is. Others have so little knowledge of this that they have a prejudice toward others who do have it. If I had to choose between my money and things or my connection to spirit, I would walk away from all my material things without question.

I certainly don't need my little angelic feelings to know X is upon us, but it's nice to know they agree. If I've studied a subject matter and have developed a certain high level of understanding, I have a way to check the validity of the information by asking my inner voice if it is correct. I have what I can only describe as a highly pleasurable direct connection to what I will call my source. At its best, chills start from the bottom of my ankles and travel all the way up my body until it feels like they are flooding off the top of my head. In a way sex is not any better than this. The most intense of these feelings are not localized in one area. In addition, these feelings or chills leave me with more energy and enthusiasm than when they started. This makes me believe that the hedonistic pursuit of pleasure is a natural thing.

Most look outside of themselves to find it. It could very well be the body was meant to create pleasure but will only last if sought after in the correct way, which might be following and finding what is true to your spirit. The many other ways of obtaining pleasure from other than spiritual, social or mental sources will eventually wear you down, sicken and eventually take you to your grave. Developing this strong connection to your source may be the key to a happy, highly pleasurable and spiritually connected life.

Now, if I'm not on the right track about what I'm questioning or pondering, I don't get any feelings. It's kind of like red light, green light. I don't know why this happens. It does work better if I keep myself in good shape physically, I've noticed. Any sort of stimulants, stress, overeating or exhaustion have a deadening effect on this process. Maybe it's the next step in man's evolution. It certainly is mine. I've painted the big picture for you. Now you will begin to see the smaller pieces fall into place almost without trying.

My Motivation

I never intended to become a writer. Somehow it became unavoidable. After finding the clarity of my understanding about Planet X, I felt that, if I didn't try to share it, I was no better than the controllers trying to hide it.

Finding the evidence for Planet X was simple and relatively easy. Now I'm going to give parts of the story that may have led me to write it. Some pieces of this story can be checked by looking at a few negligible school records. The most interesting parts cannot be verified in any way and only ring true to my spirit. I just ask you to turn them over in your mind a time or two. I will begin by saying that I believe there may be a subtle guiding hand that led me to this work.

For whatever reason, my mother taught me to read early on. I had finished my sixth grade school texts at home by the second grade. It was rather repetitious each year to receive my new school books, only to have read them years before. By the time I was 11 years old, I was an avid reader. I enjoyed a wide variety of books, from stock trading and health to spiritual works. When the school system twice suggested I skip a couple grades, my parents wouldn't allow it. By the time I was in the middle of high school, I was so bored with the education process I dropped out. A year or two later, I had talked my way into a California State University in Fresno. After a few short months, I decided that I needed to get on with my life and that the school system was not providing me the experiences I desired.

The CEO of the solar corporation I started working with at 19 had a Ph.D in social psychology. He fancied himself as more of a professor than a corporate head. Early on, after I became involved with this company, the CEO sent four us who worked with the corporation to UCLA to take a writing course and asked us to read a few books. It was kind of fun and unusual to be asked and paid to do this, and we all did our best and Aced the course.

In my mid-30's, I attended a college with the idea of becoming a computer programmer. This was no haughty Eastern university; in fact, it was only an adequate community college named Valencia in central Florida. I merrily attended the first two years of general education courses prior to my first programming class. There was something about the whole education process that was very appealing to me at that age after being away from it for a couple of decades. I've never ceased my self-education and hopefully never will.

One of the classes I especially enjoyed was a required writing course. I found ideas spontaneously came forth, sometimes, faster than I could write them down. At one point, I remember laughing out loud in class at what-all was coming through me as I cheerfully wrote it down as if copying someone else's work. I'd be one of the first to complete an in-class, timed assignment and walk away amused at my work and how easy it was to produce. The next day I'd see another grade A on my paper. Toward the end of the class semester, the teacher informed me that I should pursue a writing career. In fact she urged me to do so. I only took it as a compliment and didn't give it a second thought. Even after dropping out because of realizing I couldn't stand computer programming, I never gave writing any serious consideration.

Before I turned 40, my mother asked me to go along with her to a psychic convention. I accepted her invitation. At the convention, she paid Dolores Cannon to hypnotize me and a psychic to do a reading for me.

During the hypnotic session, I saw this wide comet-looking object passing overhead. I was atop these beautiful rolling green hills with thousands of others just standing around. The angle or trajectory of the large object overhead was such that it was obviously not going to strike Earth. I was dressed in a robe and sandals and had long blond hair. I was somehow larger also.

A giant shuttle saucer landed, and we all slowly and calmly boarded it by walking up a wide ramp. We flew off to this planet where the structures and the soil were all this same orange, brownish-red color. The dwellings all looked like they were planted and grew up out of the ground rather than built. The buildings, if you want to call them that, were like large, wide, exposed roots from so many gigantic trees. Afterwards I thought this whole vision was all my imagination. I wrote off the entire experience for a full year until I learned of Planet X. Then the significance of this session fell into place.

The psychic was next. Another hundred dollars was spent on me for this, and I was just casually going along with it to appease my mom. This psychic proceeded to tell me a whole host of personal things about my life currently, and when I was younger, only I knew. I asked her how she could know these things. She said she didn't. She explained that she was in constant communication with some sort of spiritual entity which worked with her. The information from what she called my higher self was being passed along as it came to her in real-time from her special spiritual helper. Then came the really strange part; she asked if I wanted to know about my past lives. I said, "Sure, go ahead!" She then gave me the name of who I was my very last life. She said I had been a humanitarian writer for the *Stars and Stripes* military newspaper here in the USA. In that life, I had a desire to go to Normandy beach to write first hand what hardships the troops were experiencing. Almost every time I think about this, I experience the same chills I spoke of earlier and even tears to go with them. She said I was killed there during my assignment, along with many of the soldiers. I suppose I may really have experienced first hand the worst of what the troops went through. This cursory account here may be the first chance given me to write about it since then.

She went on to say that I've been born into several lifetimes in which I was either working with or writing words. She said I was a dogmatic, bible-thumping preacher in one life. This sounds a bit embarrassing to me today.

She said I've been highly patriotic and hold the truth in high regard. In one life in early America, I supposedly had a printshop in which I worked to expose the corruption in the Railroad when it was first coming through the upper Midwest. She explained my shop was dynamited because of this, but I was not in it at the time. Maybe I've come up against these controllers before and am back on their tails again in this life exposing their underbellies.

The farthest life back we discussed was one in which I was a simple librarian at the great library of Alexandria. This, of course, was prior to the Crusades when the church decided that all such bodies of information were a threat to their existence and subsequently destroyed them.

Entrenched, corrupt institutions will never embrace knowledge that which will threaten their existence no matter how much good that truth would benefit mankind. Planet X's passing will take care of ridding the world of these institutions, and hopefully our new culture won't ever resurrect them or anything like them.

Writings directly attributed to Jesus have been found, authenticated as to their time and place, and declared blasphemy or hearsay by the church. Apparently, part of Jesus' simple philosophy was this: The entire kingdom of god can be found within each individual, and that mortar (buildings used for worship) separated man from that kingdom. This philosophy would effectively dismantle all religious structures, churches, and institutions, and thus will never be embraced by them. Their main priority is to perpetuate themselves. Anything else opposed to that, even directly attributed to Jesus, has got to be labeled blasphemy. Many other entrenched institutions follow the same *modus operandi*.

There's a simple intravenous oxygen therapy used in a clinic or two in Germany that would dissolve the AMA and cause the dismantling of the pharmaceutical companies. This oxygen therapy will never be allowed in this country. It's basically a cure-all and even rids the body of Aids and cancer quickly, simply, and inexpensively. These people who control these institutions do everything to justify and perpetuate their existence and power. Remember and understand this: disclosing the information about Planet X will not fit their objectives. It would collapse their financial institutions.

Lastly, I'd like to speculate that perhaps I volunteered or was assigned this job of shining light on Planet X in this lifetime. There's a possibility I needed to die at Normandy beach in order to show up at the right time here to do this work. Perhaps not ever obtaining a college degree was important for me to be able to shout to moon about X without be bothered by the controllers.

The Ancients had Specific Knowledge of X's Orbit and Destruction

Toward the beginning of this book, I gave you several names from ancient sources that were in fact labels given for Planet X by these people. Zecharia Sitchin's translations of ancient Sumerian text is among the best sources. His set of books will enlighten you to a past civilization that was in many ways as varied and remarkable as our own. Since civilizations come and go on every time X goes by, our history books are mainly in the dark to earth's rich and varied stories of greatness from long ago.

The 12th Planet by Zecharia Sitchin:
Landing on Planet Earth, pages 260-263, 178, and 201 — In February, 1971, the United States launched Pioneer 10. Pioneer 10 scientists attached to it an engraved aluminum plaque. It attempts to tell whoever might find the plaque that Mankind is male and female, etc., and that (Pioneer 10) is from the 3rd planet of this Sun. Our astronomy is geared to the notion that Earth is the 3rd planet, which indeed it is if one begins the count from the center of our system, the Sun. But to someone nearing our solar system from the outside, the 1st planet to be encountered would be Pluto, the 2nd Neptune, the 3rd Uranus, the 4th Saturn, the 5th Jupiter, the 6th Mars .. and the Earth would be 7th. We know today that beyond the

giant planets Jupiter and Saturn lie more major planets, Uranus and Neptune, and a small planet, Pluto. But such knowledge is quiet recent. Uranus was discovered, through the use of improved telescopes, in 1781. Neptune was pinpointed by astronomers (guided by mathematical calculations) in 1846. It became evident that Neptune was being subjected to unknown gravitational pull, and in 1930 Pluto (was located). In Assyrian times, the celestial count of a god's planet was often indicated by the appropriate number of symbols placed alongside the god's throne. Thus, a plaque depicting the god (of Saturn) placed 4 star symbols at his throne. Many cylinder seals and other graphic relics depict Mars as the 6th planet. A cylinder seal shows the god associated with Mars seated on a throne under a 6-pointed star. Ample evidence shows that Venus was depicted as an 8 pointed star. Other symbols on the seal show the Sun, much in the same manner we would depict it today; also depicted are the Moon and the cross (which is the symbol of the Planet of Crossing or the 12th Planet).

Kingship of Heaven, pages 246-248 — The (12th) Planet's periodic appearance and disappearance from Earth's view confirms the assumption of its permanence in solar orbit. In this it acts like many comets. If so, why are our astronomers not aware of the existence of this planet? The fact is that even an orbit half as long as the lower figure for (the comet) Kohoutek, (every 7,500 years), would take the 12th Planet about 6 times farther away from us than Pluto — a distance at which such a planet would not be visible from Earth. In fact, the known planets beyond Saturn were first discovered not visually but mathematically.

The Mesopotamian and biblical sources present strong evidence that the orbital period of the 12th Planet is 3,600 years. The number 3,600 was written in Sumerian as a large circle. The epithet for the planet, shar, also meant "a perfect circle" or "a completed cycle." It also meant the number 3,600. The identity of the three terms - planet / orbit / 3,600 — could not be a mere coincidence. The reign periods (a Sumarian text) gives are also perfect multiples of the 3,600 year shar. The conclusion that suggests itself is that these shars of rulership were related to the orbital period shar, 3,600 years. Kingship of Heaven, pages 242-245 All the people of the ancient world considered the periodic nearing of the 12th Planet as a sign of great upheavals, great changes, new eras. The Mesopotamian texts spoke of the planet's periodic appearance as an anticipated, predictable, and observable event. "The great planet, at his appearance dark red. "The day itself was described by the Old Testament as a time of rains, inundations, and earthquakes.

If we think of the biblical passages as referring, like their Mesopotamian counterparts, to the passage in Earth's vicinity of a large planet with a strong gravitational pull, the words of Isaiah can be plainly understood. "From a faraway land they came, from the end-point of Heaven do the Lord and his weapons of wrath come to destroy the whole Earth. Therefore will I agitate the Heaven, and Earth shall be shaken out of its place when the Lord of Hosts shall be crossing, the day of his burning wrath.

The prophet Amos explicitly predicted: "It shall come to pass on that Day, sayeth the Lord God, and I will cause the Sun to go down at noon, and I will darken the Earth in the midst of daytime. The prophet Zechariah informed the people that this phenomenon of an arresting Earth's spin around its own axis would last only one day: "And it shall come to pass on that Day that there shall be no light, uncommonly shall it freeze. And there shall be one day, known to the Lord, which there shall be neither day or night. The prophet Joel said: "The Sun shall be turned into darkness, and the Moon shall be as red blood."

Great Pyramid Dateline

The Great Pyramid Decoded, by Peter Lemesurier (1989), ISBN 1-85230-088-4, is published by Element Books Limited, Longmead, Shaftesbury, Dorset.

It gives an accurate account of the Pyramids of Giza. It contains hundreds of engineering diagrams and perhaps thousands of measurements. It's very detailed and thoroughly researched. The author is a very openminded and educated man whose own belief concurs with the views of Erik Von Daniken and Edgar Cayce as well as many Mayan and Krishna texts. He has focused upon many aspects, such as certain stones cut with "laser precision" to 1000th of an inch accuracy. The main unit of measurement used by the designer is exactly 10 millionths of the Earth's mean Polar radius. The pyramid's design base-square has sides measuring 365.242, 365.256, and 365.259 of these same units which represent the Earth's solar tropical, sidereal and anomalistic orbits. The series of datelines seem to chart the progress of certain aspects of human existence. Everything seems to "drop off the scale" around the year 2004 (though he states there is a +/- 3 year error allowance). The "Achievements of Civilization Line" drops at 1911, recovers before dropping again at 1939, recovers but drops again but less dramatically in 1967 then a bit more at 1980 and then plummets at 2004. The "Progress of materialist Humanity" Line stays fairly flat, just bobbing up and down a little all century until it hits 2004 and totally drops down off the scale in an instant!

This Is The End, My Friend — The End

I shed a few quick tears for this world and on to the next. Let me now play and be presumptuous and pretentious. Congratulate yourself for whatever modifications in your thoughts that have transpired as a result of pondering this material. Now what does that voice in you say about what you've just read? How does it speak to you? If you are any closer to finding your inner voice, these word-relayed thoughts have served a purpose. You're perhaps near enough to feel the heat and chills of knowing how to read the most important signpost in the Twilight Zone, YOUR HEART. Have you inquired inside yet for a reality check? Are you now ready to act on the information presented? The ONLY right answers are the ones you feel. They're your answers, aren't they? Some will be skeptical and without looking find confirmation everywhere they look, then prepare. Others will know it's the truth and stay in harm's way for their own perfect reasons. If this is you, keep your head high and welcome whatever comes your way. Continue to respect others, and set an example to possibly perish with the same dignity that you lived knowing a new existence awaits you after this life.

Others will ignore the reality, not move or prepare, but will somehow survive and find out the truth after it's all said and done. Some will only believe a portion. Then they will dive into the vast subject matter and be shocked to find out it's all true and much more that I've excluded for brevity and focus. Many will ignore all and perish into nothingness.

A significant thing I have discovered is that we are not our bodies and are in some way just along for the ride. I've been fully conscious outside of mine a time or two only to be swiftly pulled back in. Nudging you toward your inner voice is as important or more as planet whatever.

I'm going to live, laugh and enjoy my life before, during and after the shift while helping others along the way. I'm forewarned, informed and entertained. Now that you've read this far, you have changed forever. A part of your mind is open now that you will not be able to shut. For some of you, it will be difficult to know the reason behind many events from this point forward when loved ones around you don't care to open their eyes to the obvious. Leaving the ones you care for may not seem like an option.

Last note: Some won't have a clue as to how to look deep to find the truth of anything. Then suddenly next week, month, or year they will be doing it every moment. You'll read situations, people and events intuitively at a glance inside. In addition to this, all the while you'll be co-creating the world you want simultaneously all around you, without giving it a second thought as to why you've suddenly transformed so quickly and completely.

If it weren't for the many other groups of dedicated, tireless, persistent researchers, this book would have been much more difficult to complete. Many have gathered together mountains of evidence and are still adding to it. One of these is called Troubled Times. I tried to put together the most powerful, heavy, and glaring pieces of evidence from as many sources as possible. Go ahead and look at this one mountain for yourself it you like. It's 3000+ pages and increasing.

http://www.zetatalk.com/thub.htm (Troubled Times)

Scroll down to the bottom left hand corner of the cover page to check out the countdown clock. Click on "The Hub" and then "Scientific Data".

The rest of this book contains what I consider some of the best, most compelling, confirming material. This is only the tip of the iceberg of what I could have presented. Start with a search of Pole Shift information and see where it takes you.

In the intuitive section, I did not include many earth change prophecies and intuitives that I've read because this book is not about that. They're just a piece of the pie. Some of my ancestors were native Americans. Although none were Hopi, I have to acknowledge them for what they've done. The Hopi traditionalists have a set of prophecies, instructions and warnings. They revealed them to the public in time enough to take heed if anyone is listening. The Hopi knew about the coming earth changes and so much more that has already come to pass. *The Hopi Survival Kit* **by Thomas E. Mails fits right into everything that this book is about. They received their information around 1000 years ago and have painstakingly passed it on from generation to generation. The Kit will also open your eyes to how our government has always worked and how it continues to do so, which is a harsh reality most don't want to focus on or concede to.**

Earth Science Evidence of Abrupt Regular Earth Changes From The Approximate 3600 Year Orbit of Planet X

Earth is speaking to us loudly and clearly. Are you listening? It may be tedious to sift through, but for those who must know the archeological evidence, read on. There's volumes of more material than the studies I present here. I could have filled an entire book with it. Presenting only archeological science showing the regularity of X's passage would not give you the entire picture. You'll notice that there are references to "ice ages." This is simply a theory and not the reason for the sudden changes. The evidence points to 3500-3700 years ago, which is the best estimates some of the studies can come up with. X's passage caused many verifiable Earth changes approximately 3600 years ago. Most of the dates given correspond to the last passage or are multiples of 3600, which point to a passage several times before. When looking at the last passage for the dates for calamities one should give or take a hundred years. The science of determining dates thousands of years ago is not exact. When looking back 3 orbits ago a few hundred years one way or another should be given. The farther out in time the less accurate the dating of the changes are.

The scientists who made these discoveries did not know that they were giving evidence of Planet X's destruction from previous passages. One has to have knowledge of X's regular passage to see how these studies fit together. Rapid worldwide sea level fluctuations as much as 20 feet can be easily checked globally and the time frames verified. Now you can put the pieces together for yourself when combined with the other sources.

Scientists See Evidence of Rapid Climate Change

MSNBC *Online*, **October 28, 1999**

In a study that may sound a warning, researchers have found evidence that the world's climate can change suddenly, almost like a thermostat that clicks from cold to hot. A new technique for analyzing gases trapped in Greenland glaciers shows that an ice age that gripped the Earth for thousands of years ended abruptly 15,000 years ago when the average air temperatures soared. "There was a 16-degree abrupt warming at the end of the last ice age," said Jeffrey P. Severinghaus of the Scripps Institution of Oceanography, lead author of a study to be published Friday in the journal *Science*. "It happened within just a couple of decades. The old idea was that the temperature would change over a thousand years. But we found it was much faster." Change in Water Temperature — Severinghaus said the rapid rise in air temperature in Greenland may have been touched off by a surge in warm currents in the Atlantic Ocean that brought a melting trend to the vast ice sheet that covered the Northern Hemispherc. It still took hundreds of years for the ice to recede, but the start of the great thaw was much more sudden than scientists had once thought. This suggests, Severinghaus said, that the Earth's climate is "tippy" — prone to be stable for long periods, but then suddenly change when the conditions are right. This raises a red flag of caution.

Earth in Upheaval by Immanuel Velikovsky

(This guy was a buddy of Albert Einstein's)

The Ivory Islands, pages 4-6 — In 1797, the body of a mammoth, with flesh, skin, and hair, was found in northeastern Siberia. The flesh had the appearance of freshly frozen beef; it was edible, and wolves and sled dogs fed on it without harm. The ground must have been frozen ever since the day of their entombment; had it not been frozen, the bodies of the mammoths would have putrefied in a single summer, but they remained unspoiled for some thousands of years. In some mammoths, when discovered, even the eyeballs were still preserved. <u>(All) this shows that the cold became suddenly extreme ... and knew no relenting afterward.</u> In the stomachs and between the teeth of the mammoths were found plants and grasses that do not grow now in northern Siberia ...(but are) ... now found in southern Siberia. <u>Microscopic examination of the skin showed red blood corpuscles, which was proof, not only of a sudden death</u>, but that the death was due to suffocation either by gases or water.

Whales in the Mountains, pages 46-49 — Bones of whale have been found 440 feet above sea level, north of Lake Ontario; a skeleton of another whale was discovered in Vermont, more than 500 feet above sea level; and still another in the Montreal- Quebec area, about 600 feet above sea level. Although the Humphrey whale and beluga occasionally enter the mouth of the St. Lawrence, they do not climb hills.

Times and Dates, pages 202-203 — Careful investigation by W.A. Johnston of the Niagara River bed disclosed that the present channel was cut by the falls less than 4,000 years ago.

And equally careful investigation of the Bear River delta by Hanson showed that <u>the age of this delta was 3,600 years.</u>

The study by Claude Jones of the lakes of the Great Basin showed that these lakes, <u>remnants of larger glacial lakes, have existed only about 3,500 years.</u>

Gales obtained the same result on Owen Lake in California and also Van Winkle on Abert and Summer lakes in Oregon.

Radiocarbon analysis by Libby also indicates that <u>plants associated with extinct animals (mastodons) in Mexico are probably only 3,500 years old.</u>

Similar conclusions concerning the late survival of the Pleistocene fauna were drawn by various field workers in many parts of the American continent.

Suess and Rubin found with the help of radiocarbon analysis that in the mountains of the western United States ice advanced only 3000 years ago. The Florida fossil beds at Vero and Melbourne proved — by the artifacts found there, together with human bones and the remains of animals, many of which are extinct — that these fossil beds were deposited between 2,000 and 4,000 years ago. From observations on beaches in numerous places all over the world, Daly concluded that <u>there was a change in the ocean level, which dropped sixteen to twenty feet 3,500 years ago</u>. Kuenen and others confirmed Daly's findings with evidence derived from Europe.

Dropped Ocean Level, pages 181-183 R.A. — Daly observed that in a great many places all around the world there is a uniform emergence of the shore line of 18 to 20 feet.

In the southwest Pacific, on the islands belonging to the Samoan group but spread over two hundred miles, the same emergence is evident. Nearly halfway around the world, at St. Helena in the South Atlantic, the lava is punctuated by dry sea caves, the floors of which are covered with water-worn pebbles, now dusty because untouched by the surf. The emergence there is also 20 feet.

At the Cape of Good Hope, caves and beaches also prove recent and sensibly uniform emergence to the extent of about 20 feet. Marine terraces, indicating similar emergence, are found along the Atlantic coast from New York to the Gulf of Mexico; for at least 1,000 miles along the coast of eastern Australia; along the coasts of Brazil, southwest Africa, and many islands in the Pacific, Atlantic, and Indian Oceans. The emergence is recent as well as of the same order of magnitude, (20 feet). Judging from the condition of beaches, terraces, and caves, the emergence seems to have been simultaneous on every shore.

In (Daly's) opinion, the cause lies in the sinking of the level of all seas on the globe. Alternatively, Daly thinks it could have resulted from a deepening of the oceans or from an increase in their areas. Of special interest is the time of the change. Daly estimated the sudden drop of oceanic level to (have occurred) some 3,000 to 4,000 years ago.

Shifting Poles, pages 111, 44, and 46 — All other theories of the origin of the Ice Age having failed, there remained an avenue of approach which already early in the discussion was chosen by several geologists: a shift in the terrestrial poles. If for some reason the poles had moved, old polar ice would have moved out of the Arctic and Antarctic circles and into new regions. The glacial cover of the Ice Age could have been the polar ice cap of an earlier epoch. The continent of Antarctica is larger than Europe. It has not a single tree, not a single bush, not a single blade of grass. Very few fungi have been found. Storms of great velocity circle the Antarctic most of the year. E.H. Shackleton, during his expedition to Antarctica in 1907 found fossil wood in the sandstone. Then he discovered 7 seams of coal. The seams are each between 3 and 7 feet thick. Associated with the coal is sandstone containing coniferous wood. Spitsbergen in the Arctic Ocean is as far north from Oslo in Norway as Oslo is from Naples. Heer identified 136 species of fossil plants from Spitsbergen. Among the plants were pines, firs, spruces, and cypresses, also elms, hazels, and water lilies.

At the northernmost tip of Spitsbergen Archipelago, a bed of black and lustrous coal 25 to 30 feet thick was found. (Spitsbergen) is buried in darkness for half the year and is now almost continuously buried under snow and ice. At some time in the remote past corals grew and are still found on the entire fringe of polar North America — in Alaska, Canada, and Greenland. In later times, fig palms bloomed within the Arctic Circle.

Sea and Land Changed Places, pages 14, 74, and 180 — (Cuvier) found in the gypsum deposits in the suburbs of Paris marine limestone containing over eight hundred species of shells, all of them marine. Under this limestone there is another — fresh water — deposit formed of clay.

Much of France was once under sea; then it was land, populated by land reptiles; then it became sea again and was populated by marine animals; then it was land again, inhabited by mammals. And as it was on the site of Paris, so it was in other parts of France, and in other countries of Europe.

The Himalayas, highest mountains in the world, rise like a thousand mile long wall north of India. Many of its peaks tower over 20,000 feet, Mount Everest reaching 29,000 feet. Scientists of the nineteenth century were dismayed to find that, as high as they climbed, the rocks of the massifs yielded skeletons of marine animals, fish that swim in the ocean, and shells of mollusks. This was evidence that the Himalayas had risen from beneath the sea. In many places of the world, the seacoast shows either submerged or raised beaches. The previous surf line is seen on the rock of raised beaches; where the coast became submerged, the earlier water line is found chiseled by the surf in the rock below the present level of the sea. In the case of the Pacific coast of Chile, Charles Darwin observed that the beach must have risen 1300 feet only recently — within the period during which upraised shells have remained undecayed on the surface.

Floods — The Flood, by Charles Ginenthal

The evidence I present below is a melange of data regarding more than one global flood. Apparently, the earlier global floods occurred when major ice caps covered the continents, and later floods occurred after these were destroyed. Recent findings verify that such global floods occurred and negate the uniformitarian argument that the flood evidence indicates only local flood episodes. The basic uniformitarian argument is that the great floods were unique events caused by ice-dammed lakes unleashed when the ice dams broke. However, if individual, localized floods occurred repeatedly during the last Ice Age, they would have washed away the whale fossils found on or near the earth surface. However, whale bones and other marine fossils have been found far inland, without having been either destroyed or eroded down to tiny fragments. This strongly supports the global flood hypothesis and contradicts the local flood theory.

This evidence fully supports Velikovsky's hypothesis. If the Earth's axis tilted or the crust suddenly, violently, moved over the mantle, then the oceans would move en masse, as immense tidal waves, away from the equator and toward the poles. On the rotating Earth, due to the Coriolis force, these tidal waves would move, not only north and south, but also counterclockwise in the northern hemisphere and clockwise in the southern hemisphere. Since the Pacific Ocean lies between the continents of North America and Asia in the northern hemisphere, and the continental coastlines form an inverted V (/ \) with its apex at the Bering Strait, the tidewater would veer east, over Alaska and Canada, and west, over Asia. In the Atlantic Ocean, the tidewater would flow more easily near the poles, covering a larger area; this would create smaller continental floods. Any ice caps in these regions would be swept away from their landlocked moorings out into the northern Atlantic Ocean and would break up, depositing large amounts of detritus on the sea bed. Since neither eastern Siberia nor Alaska were covered by such a continental ice sheet, minute amounts of glacial detritus should have been deposited in the Pacific Ocean compared to that laid down in the Atlantic Ocean.

Climate Changes in Prehistory and History

Switzerland Climate Changes in Prehistory and History, By Ken Hsu
<ken@erdw.ethz.ch> — Studying the varves of Silva plana, my student Andreas Lehmann found no Holocene varves older than 4000 years, when there was no "glacial-milk" sediment..

The conclusion is inescapable: There were no varves because the Engadine lakes were not frozen every year. There were no "glacial milk" deposits when there were no Alpine glaciers! I was excited by Lehmann's discovery and called my former student Dr. Kerry Kelts at Minnesota. He headed our Limnology Laboratory at ETH-Z before accepting a professorship at University of Minnesota. Kelts was not surprised. He told me daily: "I have been telling you all those years of the 4000 BP event, and you did not listen. There was a global cooling when the Climatic Optimum came to an end.

North Africa Climate Changes in Prehistory and History, By Ken Hsu <ken@erdw.ethz.ch> — Prof. Nicola Petit-Maire, at University of Marseilles, described the vast lacustrine deposits in the Sahara desert: the sediments were laid down during a humid phase between 9,500 to 4,000 BP. Rainfall was so abundant then that Mali was not a desert but land of great lakes.

The Cro-Magnon people came across the Strait of Gibraltar from Spain to the savannas of Sahara. They hunted elephants, rhinoceros, buffaloes, hippopotamus, antelopes, and giraffes, as depicted in their wonderful rock paintings. The deserts of North Africa expanded, however, and an early clustering of cold centuries around 5200 BP caused the deterioration of environments. Hunters and grazers left Sahara and settled on as farmers of the alluvial plains of Egypt. The cooling and aridity continued and the last of the Saharan lakes dried up 4,000 BP, ending the Saharan civilization, at about the same time when the glaciers advanced in the Alps.

Mild and wet climate prevailed during the Climatic Optimum in the Near East. I visited the Canannite City Arad on the edge of the Negev Desert: it was a populous settlement of several thousand inhabitants during the Early Bronze Age. Suddenly Arad was abandoned. The deserted city showed no signs of destruction by war; the exodus was necessitated by a shortage of water supply. Indeed, the centuries-long drought in the Middle East was the cause of the collapse of the Early Bronze Age civilization in Mesopotamia, as Prof. H. Weiss of Yale and his colleagues concluded. A marked increase in aridity caused the abandonment of settlements in the north and the collapse of the Akkadian Empire in the south. The impact of it was extensive: there were synchronous collapses of the civilizations in Hindus Valley and in Egypt. The climatic catastrophe started around 2200 BC and came to an end 300 years later. This was the expression of the 4000 BP Event in Middle East.

Central Europe Climate Changes in Prehistory and History, By Ken Hsu <ken@erdw.ethz.ch> — In central Europe, the 4,000 BP Event brought, not aridity, but increased precipitation. The cold and wet climate caused the advance of the Alpine glaciers. In the region of Prealpine lakes, the Lake Dwellers had enjoyed warm and dry climate, and they had built villages on the shores of lowland lakes. When the cold and wet climate came, the settlements were flooded; the Lake Dwellers had to leave their homes when the lake-level rose. The Zurich archaeologists discovered, for example, that the villages on the shores of the lake were abandoned about 2,400 BC, and they remained uninhabited for about 800 years.

In northern Europe, cattle farming had brought prosperity to the megalithic kingdoms. The 4000 BP Event brought forth late springs and cold and wet summers. Crops were not harvested because of late planting, and cattle were famished when it became impossible to make hays. The Indo-Europeans of northern Europe had to move. Carrying battle axes and corded-ware pottery, they went to southern Russia, from there to southeastern Europe, to Anatolia, to Persia and India. and to northwest China.

China Climate Changes in Prehistory and History, By Ken Hsu <ken@erdw.ethz.ch> — The 4000 BP event hit China also. When the legendary King Huangti ruled in China, at about 3,000 BCE, mulberry trees grew in north China where elephants and rhinoceros roamed. The climate turned cold and arid then. Yu, the first king of the Xia Dynasty, received credit for having tamed devastating floods. He may in fact not have done more than his predecessors, except flooding eased when rainstorms ceased their visitation.

India — Academic Press Insight, 5 April, 1999, by Diana Steele

The people of the Harrapan-Indus civilization, who lived in what is now northwestern India, flourished between 2600 and 2000 B.C. To probe the region's climate history, a team of geologists from Israel, the United States, and India used carbon-dating and chemical analysis to examine sediments from a now-dry lake, Lunkaransar, in the Thar Desert. As the level of the briny lake fell, salts and other minerals precipitated in distinct layers. "These lake sediments give a very high-resolution record of changing lake levels, which reflect changing amounts of precipitation in the region," says Lisa Ely, a geologist at Central Washington University in Ellensburg. Ely and her colleagues found that the lake has been mostly dry for the last 5500 years. Before then, they found, the region was wet for 15 centuries — a period that ended a millennium before the Harrapan-Indus peoples began to prosper. But an arid climate by no means rules out a healthy civilization, notes Blair Kling, a historian at the University of Illinois, Urbana-Champaign. Even without plentiful rain, the Harrapan-Indus inhabitants, he says, could have depended on the Indus River for irrigation. <u>Kling says there is evidence that a flood may have forced refugees into the cities around 1600 B.C.</u>, leading to overcrowding that could have played a role in the civilization's downfall.

Sahara — In the July 15, 1999 paper published by the journal, *Geophysical Research Letters,* the Sahara desert's arid climate change occurred <u>quickly and dramatically 4000 to 3600 years ago.</u> A team of researchers headed by Martin Cluassen of Germany's Potsdam Institute for Climate Impact research analyzed computer models of climate over the past several thousand years. <u>They concluded that the change to today's desert climate in the Sahara was triggered by changes in the Earth's orbit and the tilt of Earth's axis.</u> The switch in North Africa's climate and vegetation was abrupt. In the Sahara, "we find an abrupt decrease in vegetation from a green Sahara to a desert shrub land within a few hundred years" scientists reported. No longer were grasses and other plants collecting water and releasing it back into the atmosphere; now sand baked in the stronger sun and rivers dried up. <u>The scientists do not say what caused the change in the tilt of Earth's axis.</u>

"3,600 Years Ago —The Canaanites' earliest real presence was 1550 BC (Source: *The Canaanites,* by John Grey). According to the *World Book Encyclopedia*, an unknown civilization with an alphabet that has yet to be deciphered lived in the Indus Valley (W. Pakistan). Around 1500 BC they disappeared.

Around 1500 BC, a civilization arose on the banks of the Hwang Ho river in north central China. According to *Encarta*, <u>The 1st dynasty of Babylon ended in 1595 BC.</u> In the Semitic culture, Hyksos was deposed in 1570 BC, and the Jewish exodus led by Moses happened shortly thereafter. <u>This featured a river Nile filled with "blood" and water they could not drink.</u>

The Cycladic settlement on the island of Thera was destroyed by a great volcanic eruption about 1500 BC.

Hittite internal strife caused great disorder and ended in 1525 BC with King Telipinu.

China gave birth to one of the earliest civilizations and has a recorded history that dates from some 3,500 years ago. Pottery pieces found in Fiji suggest the islands were settled in the west from Melanesia at least 3,500 years ago. Iron manufacturing originated about 3,500 years ago when iron ore was accidentally heated in the presence of charcoal.

The Tongon and Samoan islands were probably settled from Fiji about 3,500 years ago. According to M.I. Farley, author of *Early Greece*, 1970, there was total catastrophe all over Crete about 1400 BC. The Santorini eruption (about 1500 BC) was several times greater in scope than the 1883 Krakatoa eruption.

The book *Ancient Europe*, by Stuart Pigget (1965) states that around 1500 BC, Zimbabwe and Dhlodhlo were built.

According to *Earth in Upheaval*, by Velikovsky, Research by W. A. Johnston on the Niagara Riverbed disclosed that the present channel was cut by the falls less than 4000 years ago. Careful study of the Bear River delta by Hanson showed the age of this delta was 3,600 years.

A study by Claude Jones of the Great Lakes showed that these lakes have existed only 3,500 years. This is confirmed by several geographic historical maps of Michigan available in Michigan libraries. Gales obtained the same result on Owen Lake in California. Van Winkle obtained the same result on Abert and Summer lakes in Oregon.

Radiocarbon analysis by Libby also indicates that plants associated with mastodons in Mexico are probably only 3,500 years old. Similar conclusions concerning the late survival of the Pleistocene fauna were drawn by various field workers in many parts of the American continent. From observations on beaches throughout the world, Daly concluded that there was a change in the ocean level, which dropped sixteen to twenty feet 3,500 years ago. Kuenen and others confirmed Daly's findings with evidence derived from Europe.

According to Stuart Struever and Felicia Antonelli Holton, authors of the *Koster Settlement* in Koster, IL. "It is apparent that people occupied Horizon 4 for a much shorter time and less intensely than the other levels." They were referring to the site that began in 2000 BC. Other earlier sites ranged from 3900-2800 BC, and then 5000 BC.

7,200 Years Ago — According to Basil Davidson, author of *Lost Cities of Africa*, new types of humanity appeared in Africa around 5,000 BC (3500 x2). According to *Ancient Europe* by Stuart Pigget, stone using agricultural peasantry began in Europe near 5,500 BC (3750 x 2). According to a December 17, 1996 New York Times article entitled, *Black Sea Deluge May Be Tied to Spread of Farming in Europe*, an international team of geologists and oceanographers reconstructed the history of a catastrophic flood from data gathered by a Russian research ship in 1993. Seismic soundings and sediment cores revealed traces of the sea's former shorelines, showing an abrupt 500 foot rise in water levels. Radiocarbon dating of the transition from fresh water to marine organisms in the cores put the time of the event at about 7,700 years ago (5,500 BC). According to the September 10, 1996 issue of the Seattle Times: the research ship JOIDES (Joint Oceanographic Institutions for the Deep Earth Sampling) Resolution "could easily see the light colored ash deposited from the eruption of Oregon's Mount Mazama 6,950 years ago. That titanic eruption created Crater Lake and threw out at least 40 times as much magma as Mount St. Helens did in 1980 and serves as a useful marker to date mud layers. JOIDES is a Hubble telescope for the ocean, the most advanced drilling vessel in the world. "It has 12 laboratories, more than 100 research computers and can drill in water up to 27,000 feet deep." ... "The planet appears to operate in a quasi-stable mode and pops up to a new state," said NSF's Corell.

Other Cycles — According to the September 10, 1996 issue of the Seattle Times, The lodge pole pine forest suddenly died 10,900 years ago (3633 x 3). "The weather here changed so fast and so severely that the forest of the lodge pole pine that had succeeded Ice Age glaciers died in a blink." ... "This is catastrophic climate change", said paleobotanist Richard Hebda.

Ice Age glaciers retreated from the Seattle area 14,000 years ago (3500 x 4). Page 22-23 of *Early Man in the New World*, by Kenneth MacGowan (1950), shows charts of major glacial changes 18,000 years ago (3600 x 5), 25,000 years ago (3570 x 7), 40,000 years ago (3636 x 11), and 65,000 years ago (3611 x18)! According to *Encarta*, The Dalton era started about 10,500 (3500 x 3) years ago and lasted about 1,000 years in Arkansas.

The first animals used in husbandry were domesticated in southwest Asia 11,000 years ago (3636 x 3). Most sequoias suffered extinction 11,000 years ago (3636 x 3) About 11,000 years ago (3636 x 3), the axis of the earth pointed so as to give the northern hemisphere colder winters and warmer summers. Norway was inhabited 14,000 (3500 x 5) years ago. Indianapolis is located on the Tipton Till Plain, an area of flat to gently rolling land shaped 18,000 (3600 x 5) years ago. The peak of the last ice age was 22,000 (3667 x 6) years ago. The Great Salt Lake is a shallow remnant of Lake Bonneville, a large deep fresh water lake that occupied much of western Utah and parts of Nevada and Idaho from approximately 50,000 (3571 x 14) years ago to approximately 25,000 (3571 x 7) years ago. According to an October 9, 1998 article from the Associated Press and *Science* magazine, a major ice age occurred 22,000 years ago (6 x 3666).

Tsunami Signatures — From *Geo Science*, "Tsunami Along the South Coast of NSW": The first event probably occurred concomitantly with the rise of Holocene sea-level near modern levels around 7000 BP. ...The impact of these tsunami upon the coastal landscape has been profound. Several signatures provide estimates of the magnitude of run-up of these events. The height to which chaotic mixes of sediment and imbricated boulder stacks have been deposited and the height of headlands that have had a smear of clay, sand, and shell plastered across them give general estimates of the run-up height. The elevation of eroded landscape features on headlands gives information about the depth and velocity of flow. The presence of sand laminae and splayed sand units within deltaic sediments permit the landward limit of tsunami impact to be determined. This geomorphic evidence indicates that the largest tsunami waves swept sediment across the continental shelf and obtained flow depths of 15-20 m at the coastline with velocities in excess of 10 meters per second. Along cliffs, and especially at Jervis Bay, waves reached elevations of 40-100 m with evidence of flow depths in excess of 15 m. Preliminary evidence on the Shoalhaven delta indicates that waves penetrated 10 km inland for at least one event. This geomorphic evidence suggests that the New South Wales south coast is subject to tsunami waves an order of magnitude greater than that indicated by historic tide gauge records.

Recent work indicates that the southeast coast of Australia may not be the only coast to be affected by catastrophic tsunami. The geomorphic signatures of such events have been found on Lord Howe Island in the mid-Tasman Sea, along the north Queensland coast and along the northwest coast of Western Australia. At the latter location, there is good evidence that a recent wave swept more than 30 km inland, in the process topping 60 m high hills more than 2 km from the coast.

Finally, bedrock sculpturing features have been identified on the islands of Hawaii and along the east coast of Scotland. The latter location is within the zone affected by the tsunami generated by a large submarine landslide near Storegga, Norway, also 7,000 years ago.

Friday, 7 September, 2001, 18:28 GMT 19:28 UK
Giant wave hit ancient Scotland —By BBC News Online's Helen Briggs
A giant wave flooded Scotland about 7,000 years ago, a scientist revealed on Friday. The tsunami left a trail of destruction along what is now the eastern coast of the country.

It looks as if those people were happily sitting in their camp when this wave from the sea hit the camp
Professor David Smith, Coventry University
Scientists believe a landslide on the ocean floor off Storegga, south-west Norway, triggered the wave.

Speaking at the British Association Festival of Science in Glasgow, Professor David Smith said a tsunami could strike again in the area but the probability was extremely unlikely.

Radiocarbon dating of sediments taken from the coastline of eastern Scotland put the date of the event at about 5,800 BC. At the time, Britain was joined to mainland Europe by a land bridge.

Settlers at the time would have had little warning of the disaster, scientists believe. But a scattering of tools found in the sand at a hunting camp in Inverness yields some clues.

'Very destructive'

"It looks as if those people were happily sitting in their camp when this wave from the sea hit the camp," Professor Smith of the department of Geography at Coventry University told BBC News Online.

"We're talking about two, three or four large waves followed by little ones, that would have been 5-10 metres high. "These waves do strike with such force that they are very destructive," he added. "It's like being hit by an express train."

The research provides an opportunity to assess the hazard of tsunamis in more detail. They occur frequently in the Pacific Ocean due to underwater earthquakes, landslides and volcanic explosions.

Long, uncertain history

Scientists hope to find more evidence of similar past tsunamis in eastern Scotland to predict the frequency of the destructive waves. Studies of coastal sediments show that it may be possible to develop a record of past tsunamis extending back several millennia.

Dr Ted Nield, of the Geological Society of London, said: "These events have a long and uncertain time scale. "While there is no reason for mass panic, the possibility exists that the Storegga slide will go again, and it would be imprudent to ignore that fact."

Thera Eruption — The Eruption of Thera Devastation in the Mediterranean Greater Than Krakatoa — When Krakatoa exploded on August 26, 1883, it caused widespread destruction and loss of life on the coasts of Java and Sumatra. Blast waves cracked walls and broke windows up to 160 km. away. Tidal waves, reportedly up to 36 metres high, inundated the shores of the Sunda Strait, destroying nearly 300 towns and villages, and overnight more than 35,000 people lost their lives. The Changing Face of the Thera Problem Krakatoa erupted noisily. It could be heard as much as 3,000 miles away on Rodrigues Island in the Indian Ocean. Vibrations shattered shop windows 80 miles off. The energy released in the main explosion has been estimated to be equivalent to an explosion of 150 megatons of TNT.

Ships navigating the seas in the vicinity of Krakatoa reported that floating pumice in some places had formed a layer about 3 m. thick. Other shops, 160 miles off, reported that they were covered with dust three days after the end of the eruption. In fact, the dust cloud completely shrouded the area so that it was dark even 257 miles away from the epicenter. The period of darkness lasted twenty-four hours in places 130 miles distant and fifty-seven hours 50 miles away. The blackout in the immediate vicinity continued for three days and was so total that not even lamp-light could penetrate it. Stunningly beautiful sunsets were observed during the winter months in both American and Europe, thanks to the suspension of fine particles of dust in the atmosphere.

Christos G. Doumas, *Thera — Pompeii of the Ancient Aegean,* p. 141— <u>Two titanic volcanic explosions occurred in the Mediterranean in the fifteenth century BC, one on Mount Vesuvius and the other on the island of Thera near Crete. Each dwarfed the great explosion of the Krakatoa volcano in 1883.</u>

Robert Jastrow, "Hero or Heretic?," *Science Digest,* Sep/Oct '80 According to current data, the last two great eruptions of Vesuvius occurred in 3580 B.C.E and 79 C.E. (the latter being the eruption which buried Pompeiand Herculaneum). Both Krakatoa and Thera have a Volcanic Explosivity Index or VEI of 6 which rates them as "colossal" with a plume height over 25 km and a displacement volume of between 10 and 100 ks km.

Robert Jastrow, Hero or Heretic?, — Science Digest, Sep/Oct '80— Chieh Dynasty. In the twenty-ninth year of King Chieh [the last ruler of Hsia, the earliest recorded Chinese dynasty], the Sun was dimmed... King Chieh lacked virtue... the Sun was distressed... <u>During the last years of Chieh, ice formed in [summer] mornings and frosts in the sixth month [July]. Heavy rainfall toppled temples and buildings... Heaven gave severe orders. The Sun and Moon were untimely. Hot and cold weather arrived in disorder. The five cereal crops withered and died. Written during the reign of Emperor Qin c.1600 B.C.</u>

Mediterranean — *New Scientist,* 16 January 1999, p. 43 — Book Review: *Noah's Flood* by William Ryan and Walter Pitman Simon & Schuster, $25, ISBN 0684810522. There was a truly great flood around the Black Sea, recounted orally and eventually in writing by descendants of the scattered groups of survivors. Geology, climatology, archaeology, linguistics, history, and international subterfuge bordering on espionage all play a part in a fascinating story that reveals as much about how science works today as it does about the world 7000 years ago. A chance remark from a colleague set Ryan and Pitman wondering whether a similar catastrophic flood could have been witnessed and remembered as the story of Noah. <u>Strands of evidence from diverse fields slowly came together to implicate the Black Sea, around 5600 BC.</u>

Geologists Speculate on Noah's Flood Associated Press

Some biblical fundamentalists have expended great energy searching for the remains of Noah's ark. Geological research does find reason to believe there was indeed a vast, sudden and deadly flood around 5,600 B.C., close enough to the possible time of Noah to fascinate biblical literalists and liberals alike. The Ryan-Pitman candidate for the great Flood locale is what we know as the Black Sea, bordering Turkey to the north. In 1993, Ryan and Pitman joined a Russian expedition on the Black Sea and used the latest technology to examine evidence of geological patterns, soil layers and forms of aquatic life that existed in ancient times. One telltale clue: Freshwater mollusks with smashed shells gave way to saltwater creatures that had intact shells, a biological transition that could be dated through carbon-14 testing of the shell remains. From such research, the scientists spin this scenario: Until about 5600 B.C., the Black Sea was an inland freshwater lake, considerably smaller than today's saltwater sea and lying far below the level of the Mediterranean Sea.

Black Sea — "Trailing Ancient Mariners," Washington Post, September 26, 1999 — As the story is told in the Old Testament, the great flood lasted for 40 days and 40 nights, and submerged every living thing on Earth beneath 24 feet of water, sparing only Noah, his family and the pairs of animals he protected on his ark. Scientists have never found Noah or his ark, but they believe in his flood. It happened about 7,600 years ago, when the Mediterranean Sea, swollen by melted glaciers, breached a natural dam separating it from the freshwater lake known today as the Black Sea.

The theory of the Black Sea's Neolithic catastrophe was developed by Columbia University marine geologists William Ryan and Walter Pitman over three decades of research and published this year in their book *Noah's Flood*. The authors describe how the sea level worldwide began to rise as glaciers melted at the end of the last ice age 15,000 years ago. When the melt began, the Black Sea was a freshwater lake fed by rivers, among them those known today as the Danube, the Dnieper and the Don. On the lake's southern edge, a 360-foot natural dam held back the waters of what is now the Mediterranean Sea. By 7,600 years ago, sea level probably had risen to within 15 feet of the lip of the Bosporus. And then it flooded....

"For Noah's Flood, a New Wave Of Evidence," Washington Post, November 18, 1999 Scientists have discovered an ancient coastline 550 feet below the surface of the Black Sea, providing dramatic new evidence of a sudden, catastrophic flood around 7,500 years ago — the possible source of the Old Testament story of Noah. A team of deep-sea explorers this summer captured the first sonar images of a gentle berm and a sandbar submerged undisturbed for thousands of years on the sea floor.

Now, using radiocarbon dating techniques, <u>analysts have shown that the remains of freshwater mollusks subsequently dredged from the ancient beach date back 7,500 years and saltwater species begin showing up 6,900 years ago. Explorer Robert D. Ballard, who led the team that collected the shells, said the findings indicate a flood occurred sometime during the 600-year gap.</u> "What we wanted to do is prove to ourselves that it was the biblical flood, "Ballard said in an interview this week. The findings offer independent verification of a theory advanced by Columbia University geologists William Ryan and Walter Pitman that the Black Sea was created when melting glaciers raised the sea level until the sea breached a natural dam at what is now the Bosporus, the strait that separates the Mediterranean Sea from the Black Sea. <u>An apocalyptic deluge followed, inundat-</u>

ing the freshwater lake below the dam, submerging thousands of square miles of dry land, flipping the ecosystem from fresh water to salt practically overnight, and probably killing thousands of people and billions of land and sea creatures, according to Ryan and Pitman. The two scientists described the catastrophe in their book *Noah's Flood*, based on 30 years of research that began with coring samples showing the same abrupt transition from lake to sea that Ballard confirmed with his dredge. No one had ever actually seen the old shoreline, however, until Ballard's team captured sonar images of it in August.

Ryan and Pitman also that suggested that the flood may have triggered massive migrations to destinations as diverse as Egypt, western Europe and central Asia, an idea has provoked some academic controversy. Scholars also question whether any natural disaster could be conclusively identified as the inspiration for the story of Noah's flood. "All modern critical Bible scholars regard the tale of Noah as legendary," said Hershel Shanks, editor of the *Biblical Archaeology Review*. "There are other flood stories, but if you want to say the Black Sea flood is Noah's flood, who's to say no? "Shanks pointed out that biblical scholars date the writing of the Book of Genesis, from which the story of Noah is taken, at sometime between 2,900 and 2,400 years ago, and a similar event is described in the Mesopotamian Gilgamesh legend, written about 3,600 years ago. But while Ryan and Pitmando do not prove that the Black Sea flood directly inspired Gilgamesh or Noah, their theory argues persuasively that the event was probably horrific enough for scribes and minstrels to remember it for thousands of years. And regardless of the historical context, the science of the Black Sea flood stands undisputed.

Ryan and Pitman dated the event at 7,600 years ago, and they fixed the likely depth of the ancient coastline almost exactly where Ballard found it. "It feels good," Pitman said of Ballard's findings, analyzed by the Woods Hole Oceanographic Institution in Massachusetts. Pitman noted that the new research took place on the Black Sea's southern shore near the Turkish port of Synope—far from the northern waters where he and Ryan had worked. The flood, the underwater coastline, and the likelihood that ancient settlements lie on the submerged plain have added a new dimension to an already ambitious project. The region's main archaeological attraction has always been the Black Sea itself, composed mostly of dense Mediterranean salt water that immediately plunged to the bottom of the freshwater lake when the Bosporus gave way 7,500 years ago. Ever since, the less dense water on top has acted as a 500-foot-deep lid on a 7,000-foot-deep oxygen-free abyss—a watery wilderness where scientists suspect there may be 7,500 years of shipwrecks preserved in almost pristine condition.

The tantalizing prospect of exploring this environment piqued Ballard's interest several years ago. Beginning with the Titanic in 1985, Ballard has found several historic wrecks in deep water using manned submersibles and robotic vehicles. The Black Sea project, funded by the National Geographic Society and the University of Pennsylvania, began in 1995, when teams of archaeologists on land

and in shallow water began mapping Synope and its environs. Synope is about 200 miles directly south across the Black Sea's abyssal waters from the Crimea—a natural terminus for an ancient trade route.

Ballard said he intends to use a deep-sea robot next summer to look for a sea lane. "The first thing you find is trash; you didn't have Adopt-a-Highway then," he said. And where there is trash, there are sure to be wrecks. "My biggest problem is going to be trees," he added. If wooden ships can survive in the Black Sea's depths, then so can trees. The bottom could look like a forest. These difficulties, Ballard said, are different from those inherent in the search for flood-plain settlements. Many of these were probably buried—and lost forever—when a thick layer of sediment swept into the old lake with the flood waters. And Ballard suspects many others have been destroyed by the trawlers that have been scouring the sea bottom for thousands of years. Still, he said, there are plenty of "relic surfaces" near Synope, where the water simply rose quickly to submerge intact whatever lay below. Ballard's sonar sweeps this summer found a gentle coastline "frozen in time," he said. "In a perfect world you'll see a fence," Ballard said, or maybe a stockade or even a house. And there will likely be plenty of artifacts, because "when the flood came, people just had to run."

Moses' Comet

Discovering Archeology, July/August 1999, by Mike Baillie — Moses called down a host of calamities upon Egypt until the pharaoh finally freed the Israelites. Perhaps he had the help of a comet impact coupled with a volcano. A volcano destroyed the island of Santorini in the Aegean Sea (between today's Greece and Turkey) around the middle of the second millennium B.C. Researchers Val LaMarche and Kathy Hirschboeck suggest the volcano might be associated with tree-ring evidence for several years of intense cold beginning in 1627 B.C. Could that form the basis for strange meteorological phenomena recorded in the biblical book of Exodus? In the book of Exodus, which describes events a few hundred kilometers from Santorini, we read of a pillar of cloud and fire, a lingering darkness, and the parting of the Red Sea. An enormous column of ash must have hung in the sky over the eruption (the Israelites' "pillar of cloud by day and fire by night?"), and the volcano doubtless caused a tsunami, or tidal wave (which could have drowned a pharaoh's army). The Exodus story is traditionally dated to either the thirteenth or fifteenth century B.C. Those dates, however, depend ultimately on identifying the "Pharaoh of the Oppression," and historians have never proven to which ruler that infamous title referred. Many biblical scholars will disagree, but I suggest that a seventeenth-century B.C. date is not impossible.

The argument can be bolstered. Equally catastrophic meteorological conditions are recorded in the Bible for the time of King David. Psalm 18, in reference to David, speaks of terrifying events: "Earth shook and trembled. The foundations of the hills moved and were shaken. ... Smoke ... fire ... darkness ... dark waters ... thick clouds of the skies ... hailstones and coals of fire." On some chronologies, David is placed 470 years after the Exodus. The spacing between the two disastrous events recorded in Irish tree rings at 1628 and 1159 B.C. is 469 years. The Exodus story includes dust, several days of darkness, hail, dead fish, undrinkable water, cattle killed by hail, water breaking out of rocks, the earth opening, the sea parting as in a tsunami, and so on. Someone looking at the Exodus story and knowing descriptions of other distant volcanic effects might offer the possibility that the Israelites escaped from Egypt under the cover of a major natural catastrophe. There may be veiled references to comets in the biblical narrative, leading to the possibility that the Santorini eruption itself may have been triggered by a bolide (comet or asteroid) impact. David Levy, co-discoverer of the comet that bears his and Jean Shoemaker's names, has argued that the description of the "angel of the Lord in the sky over Jerusalem with a drawn sword" (1 Chronicles 21) could be a reference to a comet. The Angel of the Lord was, of course, also present at the Exodus, as it was "traveling in front of Israel's army." Further, there are indications that as the Israelites left Egypt, the night was as bright as midday. The nights over Europe were reported to have been daytime-bright after the only known modern bolide impact, the Tunguska explosion over Siberia in 1908.

These stories raise the question of whether comets recorded by the Chinese at the start and end of the Shang Dynasty, at very near the same dates, were the same as the comets that may be recorded in the Old Testament. I believe that we know the answer: In the last five millennia, several dynastic changes and dark ages have been the direct result of impacts and/or volcanoes. The consequences of such events must have been devastating, leading to apocalyptic imagery in religious writing and predictions of the end of the world. Zachariah of Mitylene lived through the environmental disaster that began about 540 A.D. In the mid-550s, he wrote in his twelve-volume records of the trials the world had survived: "In addition to all the fearful things described above, the earthquakes and famines and wars, ... there has also been fulfilled against us the curse of Moses in Deuteronomy." "The curse included pestilence, consumption, fever, fiery blasts from the skies, mildew, a rain of powder and dust, and darkness. The curse of Moses must have seemed an appropriate description of life after the impact of a piece of a comet."

Mike Baillie is a leading dendrochronologist and Professor of Palaeoecology at Queen's University, Belfast, Northern Ireland. His book, *Exodus to Arthur*, describes in detail his theory of comet encounters and turning points of civilization.

Past Cataclysms

Scottish Ocean side — Broadcasts in France are talking more and more about cataclysms. What I've noticed is that each time there is a reference to a huge tidal wave in the past which has been recently discovered. Recently, after a long speech about the Japanese tsunami, they talked about a discovery in Scotland. A huge landslide on the ocean side affected the area and all the countries around up towards Norway. They were puzzled because it's supposed to be a stable area. And this was said to have happened 7000 years ago. (2 orbits ago.)

Marine Sediment — "A 28,000 Year Marine Record of Climate Change," *Quaternary Research*, 1999, Vol. 51, No.1, pp. 83-93 — University of Bremen, Bremen, Germany: Marine sediment cores from the continental slope off mid-latitude Chile (33 degrees S) were studied with regard to grain-size distributions and clay mineral composition. The data provide a 28,000 yr. C-14 accelerator mass spectrometry-dated record of variations in the terrigenous sediment supply reflecting modifications of weathering conditions and sediment source areas in the continental hinterland. These variations can be interpreted in terms of the paleoclimatic evolution of mid-latitude Chile and are compared to existing terrestrial records. Glacial climates (28,000-18,000 cal yr. B.P.) were generally cold-humid with a cold-semiarid interval between 26,000 and 22,000 cal. yr. B.P. The deglaciation was characterized by a trend toward more arid conditions. During the middle Holocene (8000-4000 cal yr. B.P.), comparatively stable climatic conditions prevailed with increased aridity in the Coastal Range. The late Holocene (4000-0 cal yr. B.P.) was marked by more variable paleoclimates with generally more humid conditions. Variations of rain fall in mid-latitude Chile are most likely controlled by shifts of the latitudinal position of the Southern Westerlies. Compared to the Holocene, the southern westerly wind belt was located significantly farther north during the last glacial maximum. Less important variations of the latitudinal position of the Southern Westerlies also occurred on shorter time scales. (C) 1999, University of Washington.

Sea Level — "Scientists Challenge Conventional Sea Level Theory," ABC News, December 3, 1999 — Australian scientists say they have discovered evidence of rapid change in world sea levels and of a dramatic fall in geologically recent times, directly challenging current conventional wisdom. Dr Robert Baker of the University of New England, in the New South Wales country town of Armidale, has tapped the secrets of worm coatings on once-submerged rocks to shake established theory that sea levels are presently as high as they have ever been. Based on height measurements of worm coatings on rocks now well above sea level, and carbon dating tests which show them to be as recent as 3,500 years old, Baker argues that sea levels have not been steady since the last ice age, as is commonly believed. Instead, he told Australia's ABC television, it changed rapidly 3,000-5,000 years ago. "It means that the whole natural system is unstable, it's been unstable for 130,000 years." Baker and his colleagues at New England University say the sea level may have fallen quickly 3,500 years ago by as much as a meter in just 10-50 years. This means that the current rise in the sea level — normally associated with environmental warming caused by the so-called greenhouse effect — might not be that unusual, Baker said. He also said that his evidence pointed to the controversial conclusion that sea levels had once been higher than they are now.

"The conventional wisdom has been that sea levels haven't been higher. Contrary evidence was something that they weren't prepared to accept," he said. Baker's theories, which he first aired 20 years ago, were initially rejected, but are now about to receive a wider audience with their publication in the respected journal *Marine Biology*. The implications go further than greenhouse and global warming. Baker said big movements in sea levels could explain the migration of Australian Aborigines and give clues about the fate of ancient civilizations such as in Egypt.

Extinctions

According to *Discovery Magazine,* April 1999, the American Mastodon roamed here for about 4 million years until about 11,500 years ago. Another type, the Mammuthus primigenius, roamed around 400,000 years until 3,900 years ago. Both extinction times could be multiples of 3,600 years. The heyday of the woolly mammoth was the Pleistocene Epoch, stretching from 1.8 million years ago to the end of the last ice age 11,000 years ago. Mammoths thrived particularly well in Siberia, where dry grasslands once stretched for hundreds of miles, supporting a vibrant ecosystem of mammoths, bison, and other jumbo herbivores.

The mammoth fossils on Wrangell Island are the youngest that have ever been found. It was there, apparently, that mammoths made their last stand. They died out only 3,800 years ago. It had always been thought that the mammoth died out about ten thousand years ago, with the end of the ice age, but the tusk appeared to be 7,000 years old. It was so unlikely, so Buttanyan tested five more tusks, but the new dates pointed to an even more remarkable conclusion. Hidden up here [Wrangell Island] in the Arctic, the mammoth hadn't just survived the end of the ice age, it was walking these hills at the time of the Egyptian Pharaohs, only 3500 years ago. This discovery has led to the re-examination of the complex chain of "cause and effect" that made mammoths die out everywhere else, and in the process has revitalized the whole debate about how species might avoid extinction.

Magnetic Decay

Date: Mon., 10 Nov. 1997 15:06:01 GMT From: Larry Newitt <newitt@geo-lab.nrcan.gc.ca> Subject: re: Decay of the earth's magnetic field I am not familiar with the article by Barnes in the *SIS Review*, but the decrease in the earth's magnetic field to which he referred is well known. That is not to say that the strength of the magnetic field is decreasing by the same amount everywhere. Measurements of the magnetic field strength are routinely made at different places on the earth and show different rates of decrease. (The magnetic field strength is decreasing as a result of the approach of Planet X as the rotation of the Earth slows.)

Deep Quakes

http://quake.geo.berkeley.edu/cnss/catalog-search.html

An exponential increase in deep quakes down to the 500 km level since 1994, using the database provided by a private organization known as The Council of the National Seismic System, working out of Berkeley, CA, which provides earthquake data and answers questions at their web site. Note a depth down to 500 km was taken for this graph, versus a depth down to only 550 km for other pages on this site discussing this issue. This graph is thus more comprehensive, and thus more accurate. The search parameters used were: catalog = CNSS, start time = yyyy/mm/dd,hh:mm:ss, end time = yyyy/mm/dd,hh:mm:ss, minimum magnitude = 3.0, maximum magnitude = 9, minimum depth = 500, maximum depth = 700, event type = E. And the search results were:

Year,	EQ's-Avg.,	Mag.-depth
1970	113	4,35
1971	119	5,05
1972	129	4,95
1973	150	5,35
1974	130	4,9
1975	129	5,1
1976	150	4,93
1977	190	4,93
1978	152	4,94
1979	186	5,25
1980	165	5,1
1981	119	4,55

1982	126	4,93
1983	186	4,35
1984	236	4,9
1985	212	4,2
1986	247	4,7
1987	223	4,55
1988	205	4,95
1989	194	4,65
1990	223	5,4
1991	197	4,91
1992	223	4,81
1993	218	4,45
1994	229	4,9
1995	379	4,4
1996	556	4,4
1997	574	4,3
1998	499	4,56
1999	320	4,4
2000	428	4,5

2001 189 (first half only)
Look at the jump up in the # of Deep Quakes starting in 1995!
Planet X's legendary approach starts again the approximate
7 year cycle of events prior to passage.

Since 1996 a private organization known as The Council of the National Seismic System, working out of Berkeley, CA, provides earthquake data and answers questions at their web site. Their graph of earthquakes during 1996 dramatically shows the sharp increase in deep earthquakes.

Climate Changes

Arctic thunderstorms seen as latest signal of climate change OTTAWA (CP) — Canada's Inuit are seeing something unknown in their oral history — thunder and lightning. Electric storms in the High Arctic are among the evidence of climate change being reported in a new study by the Winnipeg-based International Institute for Sustainable Development. The study is believed to be the first to intensively document aboriginal knowledge of changes in the Arctic environment.

"When I was a child, I never heard thunder or saw lightning, but in the last few years we've had thunder and lightning," Rosemarie Kuptana of Sachs Harbour, NWT, said Tuesday. "The animals really don't know what to do because they've never experienced this kind of phenomenon." Researchers spent a year visiting the community of Sachs Harbour, accompanying people on their hunting and fishing trips and recording their observations on videotape. The result is a powerful portrait of environmental upheaval — melting permafrost, thinning ice, mudslides, even the disappearance of an entire lake as its once-frozen shores gave way. The freshwater fish that lived in the lake were killed as it drained into the ocean. "You used to be able to walk along the beach there;, now it's all mud," said hunter John Keogak, one of those interviewed on the video, pointing to a shoreline area. Thinner ice has made it dangerous to pursue polar bears and seals and more difficult for the bears to pursue their prey. "If this keeps up, ... the polar bears, how are they going to survive?" asks Inuit hunter Peter Esau.

Residents say the seals used to bask on ice floes in the harbour throughout the summer, but in recent summers the floes have disappeared. People now see robins and barn swallows — species that never used to come so far north. There are unfamiliar beetles and sand flies. Melting permafrost is causing buildings to tilt and has rendered roads unusable. "Climate change isn't any longer a theory, but is in fact something that's happening right now and it's affecting the lives of many of Canada's northern people," said scientist Graham Ashford. He noted the Inuit possess knowledge that can't be obtained from other sources. "The Inuvialuit hunt and trap, and they're out on the land all the time. They notice very small changes. They're telling us very clearly it wasn't like this before, and they give excellent examples of how they know that it's different...

Government Wraps Things Up By 2003

The Mars missions have the powers that be attempting to be off planet for 2003. Put this together with everything else, and perhaps you are starting now to see the big picture. I know, it's ominous. There's also much adventure for those of great faith and no fear. The passing of Planet X in 2003 is an all-encompassing, overwhelming issue. How it will impact your life, I cannot say. The decision to ignore or deal with it is yours alone. Anything can happen in the next few months. Who's to say you will be alive by the time it passes? If this is true, then learning of its existence may not be meant for you.

The White House, Office of the Press Secretary (Annapolis, Maryland). For Immediate Release, May 22, 1998. Fact Sheet Protecting America's Critical Infrastructures: PPD 63 — This Presidential Directive builds on the recommendations of the President's Commission on Critical Infrastructure Protection. In

October 1997, the Commission issued its report calling for a national effort to assure the security of the United States' increasingly vulnerable and interconnected infrastructures, such as telecommunications, banking and finance, energy, transportation, and essential government services. Presidential Decision Directive 63 is the culmination of an intense, interagency effort to evaluate those recommendations and produce a workable and innovative framework for critical infrastructure protection.

The President's policy: Sets a goal of a reliable, interconnected, and secure information system infrastructure by the year 2003, and significantly increased security to government systems by the year 2000, by: Immediately establishing a national center to warn of and respond to attacks. Ensuring the capability to protect critical infrastructures from intentional acts by 2003. Addresses the cyber and physical infrastructure vulnerabilities of the Federal government by requiring each department and agency to work to reduce its exposure to new threats; Requires the Federal government to serve as a model to the rest of the country for how infrastructure protection is to be attained; Seeks the voluntary participation of private industry to meet common goals for protecting our critical systems through public-private partnerships; Protects privacy rights and seeks to utilize market forces. It is meant to strengthen and protect the nation's economic power, not to stifle it. Seeks full participation and input from the Congress. PDD-63 sets up a new structure to deal with this important challenge: a National Coordinator whose scope will includes, not only critical infrastructures, but also foreign terrorism and threats of domestic mass destruction (including biological weapons) because attacks on the US may not come labeled in neat jurisdictional boxes; The National Infrastructure Protection Center (NIPC) at the FBI which will fuse representatives from FBI, DOD, USSS, Energy, Transportation, the Intelligence Community, and the private sector in an unprecedented attempt at information sharing among agencies in collaboration with the private sector. The NIPC will also provide the principal means of facilitating and coordinating the Federal Government's response to an incident, mitigating attacks, investigating threats and monitoring reconstitution efforts; Information Sharing and Analysis Centers (ISACs) are encouraged to be set up by the private sector in cooperation with the Federal government and modeled on the Centers for Disease Control and Prevention; A National Infrastructure Assurance Council drawn from private sector leaders and state/local officials to provide guidance to the policy formulation of a National Plan; The Critical Infrastructure Assurance Office will provide support to the National Coordinator's work with government agencies and the private sector in developing a national plan. The office will also help coordinate a national education and awareness program and legislative and public affairs.

For more detailed information on this Presidential Decision Directive, contact the Critical Infrastructure Assurance Office (703) 696-9395 for copies of the White Paper on Critical Infrastructure Protection. DOD Records Department of

Defense Records Management Task Force Semi - Annual Report January to June 1995 The mission of the Department of Defense (DOD) Records Management Task Force (RMTF) is to develop plans and draft policy to implement six strategic improvement initiatives proposed by the DOD RM Business Process Reengineering (BPR) effort completed in July, 1994 and approved by the Assistant Secretary of Defense [ASD (C3I)]. These initiatives must be implemented, with emphasis on electronic records, to reach the goal of a single Department process for managing information as records for the year 2003.

DOD Managing Information as Records: Strategic Plan — 2003, July 28, 1995 — The overall mission of records management is found in this document. The strategic plan is the Department's information management planning vehicle, which provides a broad brush perspective on purpose, vision, goals and functions which support this mission. Special attention is directed to opportunities technology offers. One foresight in understanding this mission has been expressed by: "the right information will be available to decision makers in the right format at the right time." This report proffers two simplified modern precepts. The first precept is that a record consists of information, regardless of medium, detailing the transaction of business. The second precept is that all Government employees are decision makers. DOD projects all unclassified information will be supported in a distributed electronic environment in the near future, all of which must be attended to by a standard records management process and system by the year 2003.

"Base Closings," Reuters, May 11, 1999 — Defense Secretary William Cohen and the Pentagon Joint Chiefs of Staff Tuesday urged the Senate Armed Services Committee to approve two new rounds of U.S. military base closings beginning in 2001. In separate letters to Republican Sen. John Warner of Virginia, the committee's chairman, Cohen and the nation's top military officers said more domestic bases must be closed to save money for military operations and new arms purchases. The panel is expected to vote as early as this week on a proposal to pave the way for two rounds of base closings in 2001 and 2003 when it marks up the fiscal year 2000 military spending authorization bill.

Shipbuilding Plan Overview: The cornerstone of our shipbuilding plan for the Future Years Defense Program in fiscal years 1998 through 2003 is full funding of all of the ships in the plan, including all of our submarines and the tenth and final NIMITZ class aircraft carrier, the CVN 77. Key factors used in developing our plans for the future are the number of ships now in the fleet -approximately 354 ships and submarines - and their average age. Star Wars Congressional Record, June 4, 1996 DEFEND AMERICA ACT OF 1996 — MOTION TO PROCEED (Senate — June 04, 1996) [Page: S5716] In short, our actions, if we go for and vote for the Dole star wars bill, should not be considered in a vacuum. Intended or not, implementation of the Dole star wars bill would have a far-reaching, chilling effect on the future of arms control. Often forgotten in the debate on the national missile defense is the question of whether technology is sufficiently mature enough to mandate the year 2003 as the deployment date. The record of missile interceptor testing to date and in the foreseeable future is one of more failure than success. In the rush to deploy a prototype system using highly advanced and sophisticated technology by the year 2003, we will be forsaking, Mr. President, the-fly-before-you-buy principle that has served us well in recent years. Not only will we be limiting the testing and evaluation of the system in a push to field a system at an earlier and unnecessary date, we will be locking ourselves into certain technologies which may become obsolete by the year 2003.

[Page: S5717] America's Editors Oppose New Star Wars Plans One of the most wasteful items (in the House defense budget) is the $4 billion earmarked to construct a missile defense system by 2003. This dubious "Son of Star Wars" could wind up costing as much as $54 billion before it finally could be deployed: "Fort Pork Gets Reinforced," the Miami Herald, Miami, FL, May 20, 1996. Quote from Democratic Reform News This system normally sells for sixty billion, but we're going to let you have it for five because we like you. ... The bill would order work to start on an anti-missile system (much less grandiose than the trillion-dollar Star Wars "invisible shield" President Reagan favored) that could theoretically shoot down an intercontinental missile or two launched at our territory by a small rogue country like Libya, North Korea, Iran, Iraq, or for that matter Denmark by the year 2003. Missile Defense S. 1635.

The Defend America Act of 1996, Law S. 1635, sets a clear policy to deploy by the end of thc year 2003, a National Missile Defense (NMD) system to provide a highly effective defense of the United States against the most probable source of ballistic missile threats in the post Cold War world limited, unauthorized or accidental ballistic missile attacks. The legislation does not establish a specific architecture for such a NMD system, but in order to meet the 2003 deployment date, the bill requires the Secretary of Defense to develop for deployment an affordable and operationally effective NMD system. Section 3. National Missile Defense Policy establishes U.S. missile defense policy in two areas: Deployment by the end of 2003 of an NMD system capable of providing a highly effective defense of United States territory against limited, unauthorized, or accidental ballistic missile attacks, and which will be augmented to a layered defense as larger and more sophisticated threats emerge.

Asteroid Defense Gods of The New Millennium, by Alan Alford

In 1996, The Pentagon announced a plan, sponsored by the US Air Force, to save the world by deploying missiles which would intercept "asteroids" in deep space. Politicians have indicated their intent to pass legislation which would force America to deploy this missile Defense system, code named - Clementine 2 - by AD 2003. Why the sudden haste? Is it part of the same hidden agenda that is attempting to place incredibly sensitive telescopes into deep space?

Crystal Laser Crystal Cultivator — Russian-born physicist Natalia Zaitseva has an emerald-green thumb. Using her fast-growth method, a tiny seed crystal is planted in a 6-foot rotating tank of potassium dihydrogen phosphate solution. In just six weeks, it matures into a gargantuan, 500-pound pyramid-shaped crystal. Raw crystals of that size traditionally take up to two years to grow. Zaitseva first developed her technique in Russia, but is now using it to help engineers build the world's largest laser at the US $1.2 million National Ignition Facility at Lawrence Livermore National Laboratory in Northern California. The laser, made up of 192 beams, will be housed in a complex the length of two football fields and will be used to simulate the blast of a small-scale fusion bomb and create a pebble-sized sun as hot as the real thing. But completion of the project by its scheduled 2003 launch date would be impossible without Zaitseva's fast-growth method.

Solar Maximum

The solar maximum was some time ago, I remember clearly, expected to reach its peak in its so called solar cycle in 2000. Now read this from NASA. September 23, 1998: As the Sun heads South, crossing the celestial equator today at 1:37 a.m. Eastern Time, Autumn begins for Earth's Northern Hemisphere. This Autumnal Equinox finds an increasingly active Sun steadily approaching a solar cycle maximum expected around the year 2003 (obvious disinformation). This was later returned to a solar maximum in the year 2000. (They are trying to alter maximum sun cycles from 2000 to 2003 to explain away disruptions by blaming them on the sun instead of Planet X.)

Interferometers Space Technology

3 Space Technology 3, scheduled to launch in 2003, will test technologies and flying concepts that will benefit NASA's Origins Program, which seeks answers to the origins of our universe by studying distant stars and their planets. By sending interferometers into space, NASA's goal is to image extremely distant stars, and ultimately even find and image planets like Earth around other stars!

Europa

Life May Exist On Planets In Deep Space Discovery News Brief, July 1, 1999 Life-sustaining conditions may exist on planet-like bodies in deep, interstellar space, according to a California Institute of Technology scientist. ... Lissauer points out that NASA is currently planning a year 2003 mission to Europa, one of Jupiter's moons, which is very dark, but that has an ocean thought to be composed of $H2O$. "Like Stevenson's model, life-sustaining water could exist below Europa's surface," says Lissauer. Sea Launch Hughes goes for four more ocean-platform launches CNN Interactive, June 16, 1999 Hughes Space and Communications has put in four more orders with an international venture for satellite launches from a floating platform in the equatorial Pacific Ocean, it was announced Wednesday. Sea Launch Co., a partnership between Boeing Commercial Space Co. and companies in Norway, Russian and Ukraine, already had agreements for 10 launch contracts from Hughes and five others with Loral Co. "This is a large boost in confidence from our largest customer - Hughes," said Sea Launch spokesman Terrance Scott. "It also expands our launch manifest to 2003."

Geostationary Imaging Fourier Transform Spectrometer (GIFTS)
Mission Teaming Opportunity for Geostationary Imaging Development — NASA Commerce Business Daily Issue, July 1, 1999 PSA # 2379 NASA / Langley Research Center — The Langley RC is currently conducting a study and preparing a proposal for a Geostationary Imaging Fourier Transform Spectrometer (GIFTS) mission to be operational in early 2003.

ESA Rendezvous
EUROPEAN SPACE AGENCY Project Supported by D/OPSROSETTA: ESA's Rendezvous Mission with a Comet ROSETTA represents ESA's Horizon 2000 cornerstone mission No. 3 Mission Overview The ROSETTA mission is a cometary mission, which will be launched in the year 2003 by Ariane 5. After a long cruise phase, the satellite will rendezvous with comet Wirtanen and orbit it, while taking scientific measurements. A Surface Science Package (SSP) will be landed on the comet surface to take *in situ* measurements. During the cruise phase, the satellite will be given gravity assist maneuvers once by Mars and twice by the Earth. The satellite will also take measurements in fly-bys of two asteroids.

IRIS
Astro-F (IRIS; Infrared Imaging Surveyor) The Infrared Imaging Surveyor (IRIS) is the second infrared astronomy mission of the Institute of Space and Astronautical Science (ISAS). IRIS is dedicated to infrared sky survey with much better sensitivity than that of IRAS, and is expected to add significant information on many important astrophysical problems (e.g., evolution of galaxies, formation of stars and planets, and brown dwarfs and their relation to dark matter). IRIS has a 70 cm telescope cooled to 6 K with super-fluid liquid helium and Stirling-cycle coolers. Two focal-plane instruments are installed. One is the Far-Infrared Surveyor (FIS) which will survey the entire sky in the wavelength range from 50 to 200 microns, with angular resolutions of 30 - 50 arc sec. The other focal-plane instrument is the Infrared Camera (IRC). It employs large-format detector arrays and will take deep images of selected sky regions in the near- and mid-infrared range. The field of view of the IRC is 10 arc min., and the spatial resolution is approximately 2 arc sec. IRIS was officially given a new start as the ISAS's 21st science mission" ASTRO-F" in April 1997. It is scheduled to be launched in

February, 2003 by the ISAS M-V rocket into a sun-synchronous polar orbit at an altitude of 750 km.

Nuclear Plants

Germany's Trittin Wants Speedy Nuclear Shutdown Fox News, August 21, 1999 German Environment Minister Juergen Trittin wants to shut six nuclear power plants by 2003, the news magazine *Spiegel* said on Saturday.

Firearms Registration

Canada Firearms 2003

Canadian registration of firearms by the year 2003, every legal firearm in Canada will be registered or recorded. That means every firearm in the possession of an individual or an organization, including museums, government departments and police, must be part of the national firearms registry ....

Launch Manifest

From: NASANews@hq.nasa.gov Date: Wed, 25 Jun. 1997 14:53:10 -0400 (EDT) Subj: NASA Launch Manifest is Released Msg #: 187 Debra J. Rahn Headquarters, Washington, DC June 25, 1997 (Phone: 202/358-1639)Kyle Herring Johnson Space Center, Houston, TX (Phone: 281/483-5111) NOTE TO EDITORS: N97-45 NASA LAUNCH MANIFEST IS RELEASED Copies of the NASA Mixed Fleet, Payload Flight Assignments, June 1997 edition, are available from the NASA news centers or on the Internet at URL: http://www.osf.hq.nasa.gov/manifest/ This manifest summarizes the missions planned by NASA for the Space Shuttle and Expendable Launch Vehicles (ELVs) as of June 1997. Space Shuttle and ELV missions are shown through calendar year 2003. Space Shuttle missions for calendar years 2002-2003 are under review pending the resolution of details in the assembly sequence of the International Space Station.

Mars Express

European Space Agency, Press Information Note No 22-98 Paris, France 19 June 1998 Hurry along please, for the Mars Express Any scientists wanting to send instruments to the surface of the planet Mars have until 3 July to offer a small lander that might be carried aboard the European Space Agency mission Mars Express. The selection of a lander, if any, will then be the last stage in defining the scientific payload of Mars Express, which is intended to go into orbit around the Red Planet at the end of 2003. Most judgments about Mars Express and its instruments have therefore to be made in advance if the engineers and scientists are to make sure that everything is ready for lift-off in June 2003. For more details visit the Mars Express web site European Space Agency, Press Release No. 47-98 Paris, France 5 November 1998 Mars Express wins unanimous support All fourteen national delegations in the European Space Agency's Science Programme Committee have backed the project to send a spacecraft to Mars in 2003. Development of the spacecraft will now proceed swiftly to meet the deadline of an exceptionally favourable launch window early in 2003.

Return Flight

JPL Universe, November 13, 1998 New Mars plan targets sample return Under a new plan drafted by NASA and its French, Italian and European counterparts, the consortium of space faring nations will begin development of affordable spacecraft and innovative new technologies to obtain *in situ* measurements and samples of Martian material in preparation for human exploration of the planet. The plan calls for construction of a fleet of affordable launch vehicles, orbiters, landers, rovers and Mars ascent vehicles designed to wage an all-out effort to begin returning samples of the Martian regolith as early as April 2008. ...Work on the architectural redesign began in June. Eight "tiger teams" of experts from the international scientific community, led by Elachi and Dr. Frank Jordan, manager of JPL's Mars Program Planning and Architecture Office, were formed to address issues of spacecraft design, innovative technologies and science goals for missions beginning in 2003, as well as for achieving the overall goals of the long-range Mars Surveyor Program. Recommendations were presented to NASA Administrator Daniel Goldin on Sept. 24 and, subsequently, approved for implementation. NASA will begin the series of sample-return mission in 2003.

European

"Mars Express is a mission of opportunity and we felt we just had to jump in and do it. We are convinced it will produce first-rate science", says Hans Balsiger, SPC chairman. As well as being a first for Europe in Mars exploration, Mars Express will pioneer new, cheaper ways of doing space science missions. "With a total cost of just 150 million euros, Mars Express will be the cheapest Mars mission ever undertaken", says Roger Bonnet, ESA's Director of Science. Mars Express will be launched in 2003.

Mars Surveyor Program 2003 Lander Mission

Human Exploration — NASA Commerce Business Daily Issue, June 4, 1999 PSA#2360NASA/Goddard Space Flight Center The National Aeronautics and Space Administration's (NASA's) Human Exploration and Development of Space (HEDS) Enterprise solicits proposals for investigations to be carried to the surface of Mars by the MSP 2003 Lander Mission.

Postponed

News Service, Cornell University, July 15, 1998 Astronomer confirms Cornell's new role in 2001 Mars lander mission ITHACA, NY. — Cornell University's astronomy department is working in a newly defined role on NASA's Mars Surveyor lander mission scheduled for launch in April 2001. Although the Cornell-led Athena Rover vehicle program will not be included in the mission as previously planned, "we will be doing a pretty good job of recovery" by continuing to provide most of the science for the 2001 lander, says astronomy professor Steven Squyres, the lead researcher on the project to explore and analyze part of the Martian surface. Squyres confirms that because of revised budgets and time pressures NASA has postponed the Athena Rover segment of the Surveyor Mars mission from 2001, and has tentatively rescheduled it for the 2003 Mars Surveyor launch.

Space Escape

The last page of the May 1999 issue of *Smart Money* lists Space Adventures as selling sub-orbital space travel reservations beginning in late 2002 / early 2003: Reservations are $6,000. Ticket cost is $90,000. I'm sure wealthy insiders have reserved May, 2003.

Start II

The White House Current as of: September 6, 1996 Created January 26, 1996 BACKGROUND INFORMATION: START II RATIFICATION SUMMARY START II will increase stability at significantly lower levels of nuclear weapons. Overall strategic nuclear forces will be reduced by an additional 5,000 warheads beyond the 9,000 warheads being reduced under START I. The Treaty will set equal ceilings on the number of strategic nuclear weapons that can be deployed by either side. By the year 2003, each side must have reduced its total deployed strategic nuclear warheads to 3,000-3,500. Of those, none may be on MIRVed ICBMs. Thus, all MIRVed ICBMs must be eliminated from each side's deployed forces; only ICBMs carrying a single-warhead will be allowed.

Nuclear Weapons

From Nonproliferation data site.
US - Russia
1945 1
1946 3 1
1947 5 2
1948 7 3
1949 9 4
1950 10 5
1951 40 15
1952 80 35
1953 200 65
1954 600 85
1955 2000 1000
1956 5000 1200
1957 8000 1400

1958 12000 1600
1959 15000 1800
1960 19000 2000
1961 22000 3000
1962 24000 4000
1963 26000 5000
1964 28000 6000
1965 32000 7000
1966 31000 8000
1967 30000 9000
1968 29000 10000
1969 28500 10000
1970 28000 12000
1971 28050 14500
1972 28100 16000
1973 28300 19000
1974 28600 21000
1975 29000 23000
1976 28000 24500
1977 27000 27000
1978 26000 30000
1979 25000 34000
1980 24000 37000
1981 23900 38500
1982 23700 40000
1983 23500 41000
1984 23200 42000
1985 23000 43000
1986 22000 42000
1987 21500 41000
1988 21000 40000
1989 20500 39000
1990 20000 38000
1991 19000 36000
1992 18000 34000
1993 17000 32000
1994 15000 28000
1995 13000 24000
1996 11000 20000
1997 9000 16000
1998 7000 12000
1999 5000 8000

2000 3500 3500
2001 2500 2500
2002 1500 1500
2003 500 500

Russian Subs

Nuclear Age Peace Foundation 1187 Coast Village Road, Box 123 Santa Barbara, CA 93108-2794

A former nuclear safety inspector in the defense ministry, Alexandra Nikitin raised public alarms over nuclear waste left in the Arctic by submarine accidents and haphazard disposal of spent reactor cores. "If other countries ignore this, Russia will not be the only country to suffer," he said. When one of the retired nuclear submarines begins to leak, "everything will go down to the sea, and it will be impossible to decontaminate the ocean. It will spread throughout the international fishing areas." His Report warns, "Without international cooperation and financing, a grave situation could arise which can be pictured as a Chernobyl in slow motion." ...As a consequence of his attempt to help solving Russia's environmental problems, Nikitin is facing trial for high treason and disclosure of state secrets. Nikitin was arrested, jailed for 10 months and confined to St. Petersburg for the past two years. Since then, he says, the secret service has stepped up its KGB-style harassment: repeatedly slashing the tires of his car, demanding film from his wife's camera (who had taken photos of the aggressors), keeping him awake at night with continuous prank phone calls, threatening his lawyers with violence, and more. Within the next few weeks, Nikitin is likely to be tried in secret by a judge and two laymen, who will be chosen by Russia's FSB - the KGB's successor. At this time, just over 130 nuclear powered submarines have been taken out of service and are laid up. Eighty-eight of them belong to the Northern Fleet; fifty-two still carry nuclear fuel in the reactors. Fifteen reactor compartments have been removed from the hulls and have been prepared for storage. In all probability, around 150 nuclear submarines will be taken out of service with the Russian Navy by the year 2003.

Nuclear Wastes

CNN, March 16, 1999 — The federal government could be compelled to start storing the nation's nuclear waste in Nevada far sooner than it is now prepared to do. The White House wants to focus money and effort on constructing a permanent disposal site, and thinks the thousands of tons of spent nuclear rods should remain at their respective power plants until a repository starts collecting waste in 2010. But a new bill introduced by Senate Republicans would require storing the

nation's nuclear waste at the Nevada Test Range by 2003, seven years before the White House wants to act.
Land Mines

Clinton has set a deadline to stop using land mines outside of the Korean peninsula by 2003. CNN, February 25, 1999 "We will have
destroyed all our anti-personnel landmines by the end of the year 2000 whereas the treaty sets a limit for 2003," Berlaud told reporters. 2003, ...

Biological Weapons

In 2003. Army Plans To Open Chemical Weapons Depot For Briefing (from AP) NEWPORT, Indiana (AP) — Some 1,269 tons of an oily nerve agent so lethal a few ounces could kill millions sits in steel containers among the corn and soybean fields of western Indiana while the Army works on a plan to destroy it. Officials planned to open the Indiana site to the media today for the first time in four years so reporters could attend a briefing on the military's progress and photograph the one-ton cylinders of VX nerve agent. The military doesn't expect to destroy any of the 1,269 tons of the lethal chemical weapon in Indiana until the fall of 2003, because the Army must still finish a required report on how the process may affect the environment.

Aftertimes Governments

As we get closer to 2003, a greater portion of tax dollars will be siphoned off for the supplies that the aftertimes governments will be using. You and I will not benefit from this. The bureaucrats will plan on continuing their way of life no matter what happens. There's no need to try to hunt for their well stocked facilities demanding to be fed and clothed. I'm sure there will be something like a shoot-on-site policy for stragglers that stumble on them. These bases are set up strictly for the government and only for the government. They'll probably have signs that warn of nuclear or biological contaminants. If it's true or not you won't know and won't want to find out. You don't want to be there. If they want you alive, it will only be to use you as a slave. Otherwise, you might be used for a new-fangled version of what food was made from in the beforetimes movie Soylent Green if their supplies are low. Yes, there will be cannibalism in some areas. These old government leftovers will think that they have the right to do whatever they want. Their egos will create a fantasy that convinces themselves that the surviving population will be better off in their control. I don't think so and neither should you. So stay far away from these crazy people. They'll eventually kill each other off. In the meantime, they're going to be looking for your food and supplies when they run out. They'll not be in the business of creating supplies, just using them.

Medicare

Medicare: A great debate currently rages over the Medicare Program. There is no question but that the plan will go bankrupt by 2003. And the one year increased survival that President Clinton brags about, up from 2002, is based on building an increased mortality rate into the projection tables! Obviously, federal health care planners know that whatever happens, future health care in the U.S. is NOT going to stay as good as it has been up to now! Budget Deficit Facts Entitlement's have grown rapidly. Entitlement programs combined with the government's interest payments comprised 24 percent of our budget in 1963, in 1993 they compromised 56 percent, and if present trends continue, they will comprise 69 percent by the year 2003. These projected tax rates do not show that our children are going to be taxed unmercifully; they show something worse - that our projected spending is unsustainable and that our children are being given a country that will be unable to avoid bankruptcy.

Balanced Budget

Entitlements & Handouts: A Nation of Addicts

Imagine yourself a half million dollars in debt. You don't have the option of declaring bankruptcy. What do you do? ... At this rate, by the year 2003, three-quarters of all federal spending will be "mandatory" - can't cut it. Yet we'll be much deeper in debt. So guess what's going to happen to the other one-quarter. Howard Phillips Interview on the nationally syndicated Ollie North Show on the Common Sense Radio Network on September 27, 1996... But here's going to be the trigger, Ollie. We're headed for an economic collapse of historic proportions. Last year, we paid $345 billion dollars in interest on a $5 trillion dollar debt. Neither party is prepared to even roll back spending, even freeze spending, let alone to slash it. They talk about balancing the budget in the year 2003 and they balance it by stealing money every year from the Social Security trust fund and raising revenues, increasing the amount. The dollar has been propped up because it is the reserve currency of the world. Central banks all over the world treat the dollar as if it were almost gold. Drug dealers in Columbia, black marketeers in Moscow, have dollars under their pillow because they like bucks better than they like rubles or pesos. But when the Euro (dollar) comes on line, if the central banks drop the dollar, all of those extra dollars the Fed has printed are coming home. We're going to have a massive hyper-inflationary depression and the very legitimacy of our political systems is going to be at risk. There's going to be big change, and people are going to be looking for something new.

Fuel Cells

Of the nation's 1,000 or so rural electric cooperatives to enter into a fuel cell distribution agreement, Flint Energies expects to begin selling residential systems in 2001, at a price of about $8,500. But, like others involved in the industry, Flint predicts that price will drop to less than $4,000 by 2003.

There are several other fuel cell manufacturing companies now offering a personal alternative to keeping electricity flowing when the electrical grid is gone. Do a little research; you'll find them.

Playin' Around 'N The Aftertimes

I'm gettin' in tune, right in tune, and I'm riding on you.
The word you're looking for is Creepy.
Mulder is a fictional character.
You, Planet X, and I are real.

Shake it Baby, Shake it!
Ok, now you shook it enough.
Go give it away.
The book! The book!

I know somebody you know needs to read this now.
Go! Go! Do it now!
Time is short.
Are you listening to me?
Danger, Danger, Will Robinson.

The seriousness of the situation begs for a little humor, so understand why I interject some. Ok, I'm through playing around. Here's an interesting internet site to visit to learn more about Planet X. There are lots of sights to see and sounds to listen to. Do not expect to find the depth or focus that has been presented here. It does offer great visuals.   http://xfacts.com/x1.htm

Xtra Thanks

I want to thank John Couch Adams, Urbain Le Verrier, and David Todd for determining that the perturbations or deviations of Uranus portend another planet in our solar system. Their work was done in the mid-1800's. Many others followed with the same conclusions.

Thanks goes to Percival Lowell for his insight and determination to find the 10th planet was so great, he built a small observatory in Arizona to do so about a century ago. He would have been proud to have found out that his own Lowell Observatory was one of the first PRIVATE observatories around the world to have spotted Planet X early on in 2001. I wonder what Lowell would of thought about how the coordinates were obtained? The intuitive Nancy Lieder, who claims to have telepathic communication with a group of aliens known as zetas/greys, supplied the coordinates. No matter how obtained it's the success that counts.

Thanks to Gerry Neugebauer, who was the chief scientist of the IRAS satellite. Before the media clampdown, he publicly spoke to 6 daily newspapers informing everyone they had found the 10th planet, the last day in 1983.

United States Naval Observatory astronomers Dr. Thomas C. Van Flandern and Dr. Richard Harrington, thank you very much for all your work. They had the technology at their finger tips, at the right time, and publicly announced their certainty of having found Planet X. Sometimes a martyr is needed to shine a brighter light and bring greater attention to an extremely important issue. The largest thanks again goes to Dr. Richard Harrington for paying that ultimate price. I also acknowledge the "gangsters that rule our world" crowd, for their part in creating this martyr. From their extensive mortal options menu, a so called accident was chosen, sure right.

President Vladimir Putin and Russia's Parliamentary leaders deserve a special thanks for publicly revealing what they expect to happen in 2003. They may be part of the "gangsters of our world" crowd, but they obviously don't care what the other gangsters think, say or tell them about silence! They govern their own country, and do and say what they want, when they want. I've got to admire that. I hope they'll create a better place for themselves after "the massive population shrinkage" they say is expected. Thanks goes to Andrei Shukshin for writing this story and to Reuters wire service for broadcasting it, Sept. 13, 2000.

I also want to thank the disinformation crew or media controllers for selfish reasons. They've done a great job of keeping this story out of the public's eye. The information that is available has been muddied up to the extent that most still won't consider, understand, accept or see it yet. Thanks for sacrificing millions of lives so I can still go cash checks at the bank today. I do hope to put a kink in their plans with my work here so the house of cards or financial/money/real estate markets of the world collapse a little more quickly than they've got planned. Having said that, my gut feelings tell me their plans will not be changed in the slightest and have factored myself and others like me in already. So, I'm just another proverbial fly in the ointment.

I've got to give credit to Richard W. Noone, who wrote the book *5/5/2000, Ice: The Ultimate Disaster*. He didn't quite get the correct reason for the pending calamity but only missed the time frame by 3 years. He did 20 times the research I did in areas parallel to my own. If anyone doubts in the slightest that there is not enough scientific or historical evidence of past cataclysms, you're just ignoring it. Go read his book, and while you're at it *Pole Shift*, by John White. In fact, I absolutely stand on the shoulders of John, Richard and a host of others, which I'll mention in the back pages. If you truly want to verify the science, history and regularity of these coming events, just start sifting through their reference pages to begin with.

Thanks goes out to the fumbling disinformation crew that have already changed the size, names, locations, and color of Planet X since it was first reported by them in July, 2001. They've given the blame of finding X to different observatories in May of 2001 and then went ahead and informed us they had data from 18 years ago. What color and size will it be next? Largest asteroid ever found, huh? Could you be any more obvious? Nice job!

Dolores Cannon has written several books of which I've read most. Thank you for your work Dolores. Her 3 volume set of *Conversations with Nostradamus* is the only collection of books about Nostradamus I'd consider accurate. My mother Naomi paid for Dolores to put me under hypnosis a few years ago. I arrogantly thought I couldn't be hypnotized. The hypnotic visions I described of a large comet-looking object passing overhead and thousands standing around on rolling green hilltops looking at it will forever be with me. I admit to being embarrassed about the whole experience and wouldn't talk about it for a year afterwards. Then I knew nothing of Planet X or its connection to earthchanges. I thought the whole experience was just a silly figment of my imagination at the time. So, in addition to thanks, I guess an apology is in order to Dolores. What I once thought was a complete nonsensical hypnotic session and a waste of my time, turned out to be a significant point in my awakening to Planet X.

To the ancient cultures that left so many clues, fabulous information and several descriptive names for Planet X, I bow down and salute all of you. I wonder what our culture missed, that you all discovered and knew of, which got lost in time and calamity.

Planet X and Earth deserve credit here too. Every time they meet on the fly, a whole world of evidence is created identifying the time frame of their chancy encounter. The work of many archeological crews and the hard scientific studies they've produced are too numerous to mention. These disciplined people deserve so much thanks for all their hard work in uncovering the rather obvious signs and clues over and over again.

My appreciative readers deserve my love and thanks for the confirming material they keep sending me from a wider variety of sources than even I thought existed. Reading their over-the-top reviews and thank yous' has motivated me to rewrite and add to the material. It really makes my heart sing to think I could be a cog in the big machine that helped a few.

I must lastly thank author "Mary Summer Rain" and of course "No Eyes," who was her Native American mentor, for their work together. Two books of many authored by Mary I want to mention, *Phoenix Rising* and *Daybreak*. To get a glimpse of man's peaceful future, the earth changes, and what it means to be completely connected, read *Phoenix Rising*. No Eyes could see the coming changes with perfect clarity. She may have been blind, but I can only hope to see as clearly as she did one day.

****

Additional Web sites

Cataclysm: Compelling Evidence of a Cosmic Catastrophe in 9500 B.C. — This book is a description of a massive earth disruption whose records match those of a major pole shift. This event took place approximately three cycles of 3600 years ago, or about 11,000 years ago.

http://www.knowledge.co.uk/xxx/cat/earth/

___

Did Science get it wrong? This is a description of the earth trauma which is elaborated upon in the book *Cataclysm.*

http://www.isleofavalon.co.uk/edu/g-bank/articles/phaeton.html

___

A Forest From the Past Ice Age — Trees Record the Toll of the Last Time the World Warmed. A sudden global event about 10,000 years ago served to rapidly submerge a forest of trees in Michigan.

http://abcnews.go.com/sections/science/DyeHard/dyehard000223.html

___

The Changing Face of the Thera Problem. This is a discussion of the major volcanic eruption at Thera in the Mediterranean, some 3600 years ago. This may relate to the most previous arrival of Planet X.

http://www.ucd.ie/~classics/94/Luce94.html

___

On The Possibility Of Very Rapid Shifts Of The Poles. This is an analysis which shows that impact by a modest-size asteroid could cause a very rapid shift of the poles. It is reasonable to assume that an adjacent passage by a very large planet could do the same, even though impact was not involved. The evidence is very strong for a pole shift's occurring about 11,000 years ago.

http://www.unibg.it/dmsia/dynamics/poles.html

Cyclostratigraphy. This analysis shows that the various oscillations in the rotation of the earth all have a period which is a multiple of 3600 years, the period of Planet X, which suggests that the return of this planet exerts a periodic influence on the earth.

http://www.coastvillage.com/origins/articles/pye/cyclostratigraphye/

Earth Changes from Two Sources. This page includes a summary of the upcoming earth changes from two sources: (1) An analysis of messages received from spiritual sources during many Near Death Experiences, and (2) a commentary on these messages from spirit guides who are speaking through a gifted medium.

http://www.afterlife101.com/experiences2/research2_6.html

A Talk with Ruth Montgomery. The first intuitive I chose to give a synopsis of was Ruth Montgomery. In her lifetime, she was not just a prophet but a former White House journalist. She provides messages from her spirit guides relating to the afterlife and also about upcoming earth changes.

http://www.geocities.com/HotSprings/Spa/2366/montgomery.html

For preparedness information contact
 http://www.survivalcenter.com
360-458-6778

More information pointing to 2003:
http://www.astrosite.com/__JanM3.htm

This is important late-breaking-news!

Until this time, Zecharia Sitchin had yet to go on record to state Nibiru/Planet X would be arriving anytime soon. *Nexus Magazine,* October-November, 2001 issue,

Page 68: World renowned scholar and archeologist Zecharia Sitchin states, "There is one more planet in our solar system, **not light years away**, that comes between Mars and Jupiter every 3,600 years. I prophesies the return of this plant, **called Nibiru, <u>at this time</u>."**

I've done independent investigation to confirm that this was **not misread or quoted incorrectly.** Sitchin' does make these kind of statements occassionally by not consistently.

Inspired by "Blindsided" Di, an appreciative reader, created this egroup for regular updates, PX information, and preparedness issues. You're invited:

http://groups.yahoo.com/group/preparation2003

If you decide to join, my strong suggestion is to choose the "no-email" option. It's much easier and faster to
bookmark the main page and go there to read the messages of your subject choice, on-site, at your convenience, than to have your email box fill with emails. The following are samples of some interesting messages that came through.

This first message is from an avid sportsman in Colorado.

"I don't know whether or not I am happy to bring this to you guys and dolls. I spent around 138 bucks on a power zoom scope from Technoscout.com, better known as the Yukon scope.

Well, to make a long story short, I have found a very interesting object in the area given to me by zetatalk.com. One thing I have to say is Holy (you know what!) Keep looking towards those coordinates. You're gonna get a big surprise. I certainly did by the reaction in my throat.

I strongly believe by what's going on in the world news that this just could be a distraction, to the bigger picture, get my drift!

Right now I live in an area known as Copper Mountain, CO. Boy, the days of bad ski reports pales in comparison. The bigger picture is well on its way. As I say to everyone, don't believe me, believe your own truth.

I sure hate to be the one to say I hate it when I'm right. Prepare yourselves, this is a ticket that's priceless at the box office.

See ya soon!"

This next message is from an appreciative reader in San Diego, CA.

"To those who are wondering if this (Planet X) is real or not and are brave enough to be sure, I have written a short essay about how I crossed over the threshold into the land of knowing yesterday. But don't read on until you are sure that you really don't want to stay in the land of maybe. It will definitely change your life—-for example, right now I am up at 3:00 A.M. because I couldn't go back to sleep! Of course, what I say may not convince you, but may just be one thing to prod you on in your quest. Each of us will have a different thing that gives us that undeniable 'Aha!' I have always been the kind of person that likes to know the truth, feeling that knowledge is bliss, instead of ignorance. I read Mark's book, "Blindsided, Planet X, Earth Changes". But, I really longed to talk to someone in the know firsthand. I also spent up to seven hours a day on the Internet over the four-day holiday looking up Planet X on all the search engines. I disregarded all the channeled information because it is too contradictory, especially about the year, and I wanted to know when this is happening so that I could live as though it were my last year, or plan to survive. I prayed to know the truth.

Then I got the idea to call dome builders and ask them about the dome prices, etc. I then asked if they had heard about Planet X. One talked to me for about half an hour. Yes, in fact, he knew quite a bit. I have promised to keep his anonymity from the Internet so I cannot give away his identity but what he shared with me will hopefully give you some answers. He said some scientists told him about it and he called some friends of his that he was in the military with years ago. He trusted them because they had become scientists in NASA.

What they went on to describe is why I can't sleep at 3:00 A.M. Two things are certain, they told him: 1) Planet X exists. 2) It is coming between May 15 and May 30 of 2003. The rest is speculation. But, if it's good enough for the NASA scientists, it's good enough for me. The earth will stand still for about three days and then, in one hour, rotate a full 90 degrees (the pole shift) during which time winds will be an average of 200 miles per hour. Every volcano on earth will erupt and of course, there will be many earthquakes, so two thirds of earth's population will die in that one-hour surprise. Then another 20% will starve to death during the next six months because the volcanic ash will cover the earth and keep out sunlight for six months. About 10% of the population (about 600 million) will survive. He said NASA's top scientists wanted to go to Mars and survive but that didn't work out. So they are making a space station, which will be able to maneuver itself on the other side of earth, using earth as a shield between the space station and Planet X. This way the top scientists will live. He said it is already affecting Neptune's gravitational pole and once it gets close to Pluto, it will take only three months to get here. He confirmed what Mark said (although he never heard of this book), which is that Planet X goes very slowly and then very, very fast all of a sudden. But, once near Pluto it will be a red dot in the sky, which will gradually get bigger during several months, and can be seen with the naked eye six weeks before the pole shift. He confirmed that the northern inland part of the States is good; it is where the scientists are going.

This will be near the equator after the shift. After hearing all this, I wonder whether I want to survive. I don't relish the idea of seeing starving people. I haven't made up my mind, but it is definitely shifting me into an altered state just knowing this. I am living my life fully appreciating everything, which began after reading Mark's book and is now intensified. For those of you who are having second thoughts about survival, I cannot recommend highly enough Dr. Michael Newton's research on the afterlife, "Journey of Souls" and "Destiny of Souls". Read those if you have any fear of death.

If you want to survive 200 mile per hour winds, though, you need a dome house and lots of food stockpiled! Another thing: I have known about the secret elite government for about three years. But after finding out the truth, I feel that even if we had a benevolent government, they are wise not to publicize this. They are powerless to find the resources to save everyone. Besides, it is wise to let the masses enjoy their lives till the last. I am now working on forgiving them for all they have done to mess up this planet; how this all plays out may actually make their doings a moot point. Hopefully, as this man quoted his friends, Planet X will be the great equalizer, dethroning the powers that be."

This last message was a follow-up to a call I received from a well-to-do businessman in Germany.

"Dear Mr. Hazlewood, Thank you very much for our telephone conversation this past Saturday. As I explained to you based on a German scientists request, I was asked to remote view an Edgar Cayce Earth change map. The Professor in question's method allows an individual to enter a highly alert state of 'extended awareness' very effortlessly within some minutes using a special frequency device. I was astounded at what I saw after I was asked to 'view' and detail this Earth change map. The detail is fairly lengthy as I described to you. But the amazement continued when, upon my return from France 3 days after this viewing session, I spoke to a friend about my experiences and he told me that he had visited your website and I should also look into it based on my experiences with the remote viewing of the oncoming planet. I basically read on your website the confirmation of the remote viewing experiences I had.

I am not given to jumping to conclusions and accepting doomsday predictions. So I telephoned the Professor and he told me that he was at a meeting in Russia last year and fellow scientists let it be known that, indeed, out of one of Russia's largest observatories was observed a "new" planet approaching Earth.

We will now try to get additional data and photos through the Professors sources as well as 2 other potential observatories that may help further substantiate this unbelievable scenario. Please keep us informed of any additional "solid" data you may receive. I certainly feel a sense of urgency given the time line but want to be sure that all is verified scientifically as well, as I do not intend to make the major life changes necessary just based on hype or delusions. Having said this I also feel that what was remote viewed was clear and specific and very much in line with what you have stated in your document as well as some additional data I have provided you with. I would also ask you to respect my privacy at this point due to some public profile I have here in Europe. I will verify what sources of information may be available to us and report back to you within 3 weeks. If the data proves out I will then probably take a more public stance.

With kind regards, BJ"

Mark Hazelwood
Order Eastern Business hours M-F 1-888-707-7634
http://metatech.org/plant_x_nibiru_earth_changes.html
http://www.prep2003.com
http://www.survivalcenter.com/Planetx.html
http://groups.yahoo.com/group/planetx2003
For Updates See:
http://www.planetx2003.com